LANDFORMS and TOPOGRAPHIC MAPS

Illustrating Landforms of the Continental United States

WILLIAM BAYLY UPTON, JR.

Member of the Association of American Geographers
and the American Congress on Surveying and Mapping

John Wiley and Sons, Inc.　　New York　　London　　Sydney　　Toronto

Copyright © 1970 by John Wiley & Sons, Inc.

All rights reserved. No part of this book may be reproduced by any means, nor transmitted, nor translated into a machine language without the written permission of the publisher.

Library of Congress Catalog Card Number: 73-107592
SBN 471 89642 X

Printed in the United States of America

10 9 8 7 6 5 4 3 2 1

LANDFORMS and TOPOGRAPHIC MAPS

PREFACE

It is almost impossible to visualize or appreciate the landforms of any area without the aid of topographic maps. They give vivid details of the physiographic features of the surface geology of the areas they represent. Because of this advantage, they can be invaluable to the beginning geology or physical geography student.

I have long wanted to prepare a book containing some of the more significant topographic maps, with some accompanying textual description of each. This book is a result of that desire. I have planned it to be short, interesting, and pertinent to all sections of the United States. It is an outgrowth of the USGS *Portfolio of One Hundred Maps*, a project which I organized. The *Portfolio* has been used for several years in many earth science classrooms. The landforms in the *Portfolio* are highly authentic, since they were selected by some sixty well-known geologists, each very familiar with the section of the country to which he was assigned.

This book differs from the *Portfolio* in that it is much smaller, very much less expensive, and highly portable, containing only portions of the original quadrangles. The result, it is hoped, is a publication which students can take home after classes in geology or physical geography. All the text material is presented as clearly and simply as possible so that it can be read and understood with no loss of time.

I hope that my efforts to eliminate all errors have been successful. If some have crept by, I also hope that readers will report them to the publisher so that future editions can be improved.

I must add my sincere gratitude to the many wonderful people whose knowledge, unselfishly contributed, made this the convenient and accurate work that it is, beginning with the Geological Survey *Portfolio of One Hundred Topographic Maps Illustrating Physiographic Features*, the forerunner of *Landforms and Topographic Maps*.

Mountain View, California
January 1970

William Bayly Upton, Jr.

This earth science book is dedicated to the many wonderful people, too many to list, who helped make this an accurate and interesting study of maps and the land.

CONTENTS

Introduction	1
How to Use the Book	3
Physiographic Features (Landforms)	5
Location of Quadrangles	14
Maps and Descriptive Pages	
1. Alabama, Mobile	17
2. Arizona, Antelope Peak	19
3. Arizona, Bright Angel	21
4. Arkansas, Waldron	23
5. California, Furnace Creek	25
6. California, Point Reyes	27
7. Colorado, Juanita Arch	29
8. Connecticut, New Britain	31
9. Delaware, Little Creek	33
10. District of Columbia, Washington West	35
11. Florida, Jacksonville Beach	37
12. Georgia, Warm Springs	39
13. Idaho, Menan Buttes	41
14. Idaho, Thousand Springs	43
15. Illinois, Effingham	45
16. Indiana, Oolitic	47
17. Kentucky, Mammoth Cave	49
18. Louisiana, Campti	51
19. Maine, Katahdin	53
20. Maine, Mount Desert	55
21. Maryland, Cumberland	57
22. Massachusetts, Lynn	59
23. Massachusetts, Provincetown	61
24. Michigan, Fennville	63
25. Michigan, Jackson	65
26. Minnesota, Virginia	67
27. Mississippi, Philipp	69
28. Missouri, Ironton	71
29. Montana, Chief Mountain	73
30. Montana, Ennis	75
31. Nebraska, Ashby	77
32. Nevada, Sonoma Range	79
33. New Hampshire, Monadnock	81
34. New Mexico, Ship Rock	83
35. New York, Catskill	85
36. North Dakota, Pelican Lake	87
37. Ohio, Maumee	89
38. Oregon, Crater Lake	91
39. Pennsylvania, Altoona	93
40. Pennsylvania, Tyrone	95
41. Rhode Island, Kingston	97
42. South Carolina, Mullins	99

43.	South Dakota, Sheep Mountain Table	101
44.	Texas, East Brownsville	103
45.	Texas, Guadalupe Peak	105
46.	Utah, Jordan Narrows	107
47.	Vermont, Brandon	109
48.	Virginia, Strasburg	111
49.	Washington, Mount Rainier	113
50.	Wyoming, Mount Bonneville	115

Geologic Maps Covering Topographic Quadrangles
List 1 U.S. Geological Survey, Geologic Maps Published Separately 117
List 2 U.S. Geological Survey, Book Publications Containing Geologic Maps 117
List 3 Available State Publications 119
List 4 Publications of Professional Societies 120
List 5 Other Publications 121
Bibliography 123
Glossary 125
Topographic Map Symbols 135

LANDFORMS and TOPOGRAPHIC MAPS

INTRODUCTION

The continental United States contains a great variety of landforms which offer dramatic contrasts to the student or cross-country traveler. Mountains and deserts, glaciers, and areas of deep canyons and broad plains are examples of the nation's varied surface.

What is a landform? A landform or physiographic feature is one of the multitudinous features that make up the surface of the Earth. Although the Moon has a few landforms, they are different because there is no water for shaping them as there is on Earth.

Exactly what is a topographic map? This question requires considerable explanation because the contents of a topographic map are very different from those of most other maps such as the familiar road map. The reader of this book must thoroughly understand how to use topographic maps. The key is the word "contour." On the map, the contour lines are shown in brown. Imagination and practice are requirements for understanding these contour lines. To visualize hills, valleys, and the many other landforms, it is essential to read the features of the land as portrayed by contours. As soon as this is accomplished, the dimension of height and depth makes the visualization of the landforms a reality.

To understand the contour line, think of it as an imaginary line on the ground which takes any shape necessary to maintain a constant elevation above sea level. For example, the shoreline of the sea is in effect a contour representing the zero elevation or sea level. If the sea should rise upon the land, the new shoreline would be a new contour line. All successive rises would be contours, and a vertical difference between them may then be established.

For uniformity, it was decided long ago to establish a contour difference or interval in feet. A contour interval and scale are determined and established for each topographic map or group of maps, depending largely upon the flatness or steepness of the land of the particular area being mapped. The primary needs for the map are also considered. In the flat or lowlands, the interval can be as small as 1 foot. But it may become 5, 10, 20, 40, 50, 80, or 100 feet as the steepness of a hilly or mountainous land increases.

Several of the features depicted on a topographic map are illustrated on the accompanying bird's-eye view of a river valley and adjoining hills. The river flows into a bay which is partly enclosed by a sand spit. On both sides of the valley are terraces through which streams have cut gullies. The hill on the right has a smoothly eroded form with shallow streams and gradual slopes above a wave-cut cliff. The hill on the left descends from a steep bluff; it then falls off gently and forms an inclined plain crossed by a few shallow gullies. An improved road follows the shore of the sea; a dirt road provides access to a church and two houses.

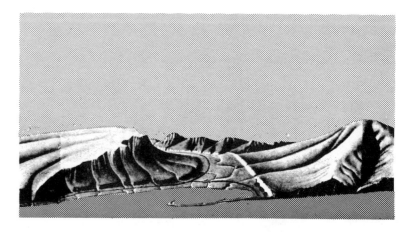

The lower part of the illustration is a topographic map representing the same contours, drainage lines, and shorelines all projecting down directly beneath the features on the landscape sketch. Elevations are represented by contour lines. The vertical difference between contours in

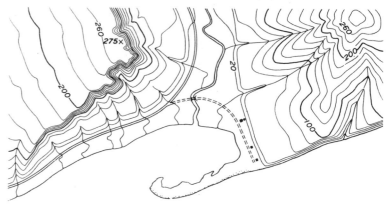

this illustration is 20 feet, this being considered the correct interval for readily representing the shape of the landscape.

From the terrace on the right, the well-separated contours continue up the gradual slope to the 280-foot contour which encircles the rounded summit. In contrast, the contours ascending from the terrace on the left abruptly become very close together near the summit where six very close contours indicate an escarpment 120 feet high. Another steep area adjoining the sea on the right side of the map has five contours very close together, indicating a wave-cut cliff 100 feet high.

As a convenience for the map users, an index contour, generally every fourth one, is accentuated by a heavier line which has its altitude marked in several places. Occasionally an intermediate contour is marked with its altitude.

The official adoption of the topographic map for our country dates back to 1879 when Congress established the Geological Survey. This Bureau was given the responsibility of planning and completing the detailed mapping. Approximately 70% of the task has been accomplished. Priority for selection of areas follows requests from federal government agencies as well as from the states. High standards of accuracy have been maintained throughout the program. Moreover, most nations of the world have been mapped or are in the process of completing their national coverages.

Topographic maps have many uses as fundamental tools for planning and executing projects that are necessary to our modern way of life. They are of prime importance in planning airports, highways, dams, pipelines, transmission lines, industrial plants, and countless other types of construction. Topographic maps are an essential part of geologic and hydrologic research, of mineral investigations, and of studies on the quantity and quality of water. They greatly facilitate the study and application of flood control, soil conservation, and reforestation. Intelligent and efficient development of our natural resources depends on the availability of adequate topographic maps. Familiarity with topographic maps can be a valuable addition to one's storehouse of knowledge.

HOW TO USE THE BOOK

1. After turning to a map, select a physiographic feature (physiographic feature and landform are similar terms; see Glossary) from the list on the page opposite and then look up its definition in the Glossary. A colored star (✻) next to a feature on the list indicates that it is especially well represented on the corresponding map.

2. Identify the feature as defined, following the clues given on the list.

3. Examine the feature, using the contours for measuring height or depth and a ruler or scale for measuring horizontal dimensions.

4. Continue your examination of features, consulting the Glossary for terms. Perhaps some words will be strange to you until you become familiar with them by repeated use. Many features are shown on several different quadrangles.

5. Refer to the "Physical Divisions of the United States" map on the back of the Geological Survey *Index* for the specific location of each map within its province and section.

LIST OF PHYSIOGRAPHIC FEATURES (LANDFORMS)*

COASTAL FEATURES AND SHORELINES

Abandoned shoreline (Pleistocene)
Mobile, Ala.

Ancient beach ridges
Jacksonville Beach, Fla.

Ancient cuspate bar, spit, and tombolo
Jordan Narrows, Utah

Barrier beach
Jacksonville Beach, Fla.
Point Reyes, Calif.

Battered sea cliff
Point Reyes, Calif.

Bay
Little Creek, Del.
Lynn, Mass.
Mobile, Ala.

Bay, Carolina
(*See* Miscellaneous Features)

Bayhead bar
Lynn, Mass.

Baymouth bar
Kingston, R.I.
Lynn, Mass.
Provincetown, Mass.

Bayou
Campti, La.
Philipp, Miss.

Beach
Kingston, R.I.
Jacksonville Beach, Fla.

Bonneville and Provo shorelines
Jordan Narrows, Utah

Cape
Point Reyes, Calif.
Provincetown, Mass.

Coastal bars
Jacksonville Beach, Fla.

Coastal terrace
(*See* Pamlico)

Cove
Mount Desert, Me.

Delta
East Brownsville, Tex.

Deltaic channels
Mobile, Ala.

Distributary channels
Mobile, Ala.

Distributary stream on delta
Point Reyes, Calif.

Drowned coastline
Mount Desert, Me.

Drowned river
Catskill, N.Y.
Washington West, D.C., Va.

Drowned valley
Mobile, Ala.
Point Reyes, Calif.

Estuary
Point Reyes, Calif.
Washington West, D.C., Va.

Fiord
Mount Desert, Me.

Headland, truncated
Point Reyes, Calif.

Lagoon
Point Reyes, Calif.
Provincetown, Mass.

Marine terrace
Mobile, Ala.

Neck
Kingston, R.I.

Pamlico shoreline
Jacksonville Beach, Fla.
Mobile, Ala.

Pamlico terrace
Little Creek, Del.
Mobile, Ala.

Point
Point Reyes, Calif.
(*See* Plateau Features)

Prograded shore
Jacksonville Beach, Fla.

Raised beach ridges
Fennville, Mich.
Jacksonville Beach, Fla.

*Arranged by types with the names of the quadrangles on which the features are most clearly shown.

COASTAL FEATURES AND SHORELINES — Continued

Raised spit and hook
Fennville, Mich.

Sand spit
Point Reyes, Calif.

Sea stacks
Point Reyes, Calif.

Silver Bluff beach, lagoon, and shoreline
Jacksonville Beach, Fla.

Silver Bluff terrace
Little Creek, Del.

Spit, compound recurved
Provincetown, Mass.

Tidal marsh or swamp
Little Creek, Del.
Lynn, Mass.

Tombolo
Lynn, Mass.

Wave-cut cliff
Point Reyes, Calif.

ESCARPMENT FEATURES (cliffs, cuestas, hogbacks, etc.)

Allegheny front
Altoona, Pa.
Cumberland, Md., W. Va.

Ancient river bluff
Philipp, Miss.

Cliff
Bright Angel, Ariz.
Guadalupe Peak, Tex.

Cuesta
Mammoth Cave, Ky.

Escarpment
Juanita Arch, Colo.

Faceted river bluffs
Washington West, D.C., Va.

Facets
Furnace Creek, Calif.

Faultline scarp
Furnace Creek, Calif.

Fault scarp
Guadalupe Peak, Tex.

River bluff
Maumee, Ohio

West front of Green Mountains
Brandon, Vt.

GLACIATION FEATURES FORMED BY ALPINE GLACIATION

Arete
Chief Mountain, Mont.

Biscuit-board topography
Katahdin, Me.

Cirque
Chief Mountain, Mont.
Mount Bonneville, Wyo.

Cirque lake
Chief Mountain, Mont.

Col
Chief Mountain, Mont.

Compound cirque
Mount Bonneville, Wyo.

Cyclopean stairs
Chief Mountain, Mont.

Finger lakes
Chief Mountain, Mont.
(See Glaciation Features, Continental)

Glacial trough
Chief Mountain, Mont.

Glacier
Chief Mountain, Mont.
Mount Rainier, Wash.

Hanging valley
Chief Mountain, Mont.
Mount Rainier, Wash.
(See Valley Features)

Matterhorn
Chief Mountain, Mont.
Mount Bonneville, Wyo.

Medial moraine
Mount Rainier, Wash.

Nunatak
Mount Rainier, Wash.

Pater Noster lakes
Mount Bonneville, Wyo.

Subsummit erosion surface scoured by ice cap
Mount Bonneville, Wyo.

Tarn
Mount Bonneville, Wyo.

U-shaped valley
Chief Mountain, Mont.

GLACIATION FEATURES RESULTING FROM CONTINENTAL GLACIATION

Abandoned glacial channels
Jackson, Mich.

Abraded bedrock hills
Brandon, Vt.
Monadnock, N.H.

Bedrock knobs
Catskill, N.Y.
Lynn, Mass.

Coteau du Missouri
Pelican Lake, N. Dak.

Deranged drainage
Monadnock, N.H.

Dissected glaciated plateau
Catskill, N.Y.

Drainage diversion, glacial
New Britain, Conn.

Drainage reversal, glacial
New Britain, Conn.

Drumloidal hills
Kingston, R.I.

End moraine
Kingston, R.I.

Esker
Jackson, Mich.

Finger lake
Chief Mountain, Mont.

Glacial drift
Brandon, Vt.

Glacial Lake Maumee
Maumee, Ohio

Glacially rounded hills
Virginia, Minn.

Ice-carved strike ridges
Catskill, N.Y.

Intermorainal lowland
Fennville, Mich.

Kalamazoo moraine
Jackson, Mich.

Kames
Jackson, Mich.

Kames and kettles
Kingston, R.I.

Kettle holes
Monadnock, N.H.

Kettle (with lake or pond)
Jackson, Mich.
Pelican Lake, N. Dak.

Knobs and kettles
Jackson, Mich.

Lake border moraine
Fennville, Mich.

Lakes and ponds in glacially scoured bedrock basins
Lynn, Mass.
Mount Desert, Me.

Lobate washboard moraine
Pelican Lake, N. Dak.

Morainal lakes
Monadnock, N.H.

Morainic topography
Jackson, Mich.

Mountains and islands modified by glaciation
Mount Desert, Me.

Obstructed drainage
Brandon, Vt.

Outwash-filled channels
Kingston, R.I.

Outwash terrace
New Britain, Conn.

Pitted outwash plain
Jackson, Mich.
Kingston, R.I.

Ponds in kettles
Jackson, Mich.

Poorly integrated drainage
Jackson, Mich.
Pelican Lake, N. Dak.

Spillway from Glacial Lake Hartford
New Britain, Conn.

Swell and swale topography
Pelican Lake, N. Dak.

Terminal moraine
Jackson, Mich.

Valparaiso moraine
Fennville, Mich.

MISCELLANEOUS FEATURES

Badlands
(*See* Plateau Features)

Carolina bay
Mullins, S.C.

Continental Divide
Chief Mountain, Mont.
Mount Bonneville, Wyo.

MISCELLANEOUS FEATURES — Continued

Dam and reservoir
Jordan Narrows, Utah

Fall line
Washington West, D.C., Va.

Fine-textured topography
Sheep Mountain Table, S. Dak.

Great Raft
Campti, La.

Highest point in Texas
Guadalupe Peak, Tex.

High relief topography
Mount Rainier, Wash.

International boundary (channel of Rio Grande)
East Brownsville, Tex.

Intracoastal waterway
Jacksonville Beach, Fla.

Laurentian Divide
Virginia, Minn.

Low relief
East Brownsville, Tex.

Lowest elevation in United States
Furnace Creek, Calif.

Natural bridge
Furnace Creek, Calif.
Juanita Arch, Colo.
Mount Rainier, Wash.

Rock sculpture controlled by fractures
Mount Desert, Me.

Sea level and below sea level contours
Furnace Creek, Calif.

Underwater contours (depth curves)
Point Reyes, Calif.

MOUNTAIN FEATURES (ridges, hills, faults, folds, etc.)

Accordant summits
Strasburg, Va.

Anticline
Altoona, Pa.

Anticlinal ridge
Cumberland, Md., W. Va.

Canoe-shaped mountain
Strasburg, Va.
Tyrone, Pa.

Concave slope
Monadnock, N.H.

Dissected block mountains
Furnace Creek, Calif.
Sonoma Range, Nev.

Dissected upland
Ironton, Mo.

Esplanade
Juanita Arch, Colo.

Fault block mountains
New Britain, Conn.

Faultline valley
(*See* Valley Features)

Flatirons
Ship Rock, N. Mex.

Folded mountains
Waldron, Ark.

Folds en echelon
Waldron, Ark.

Hogback
Cumberland, Md., W. Va.
Tyrone, Pa.

Hogback — Continued
Waldron, Ark.

Island mountains (inselbergs)
Antelope Peak, Ariz.

Isolated ranges
Sonoma Range, Nev.

Klippe
Chief Mountain, Mont.

Matterhorn
(*See* Glaciation Features)

Migrating divide
Bright Angel, Ariz.

Monadnock
Monadnock, N.H.

Mountain peak, isolated
Monadnock, N.H.
Mount Rainier, Wash.

Pass
Chief Mountain, Mont.

Ridges formed of folded hard strata
Waldron, Ark.

San Andreas Rift
Point Reyes, Calif.

S-shaped ridge
Strasburg, Va.

Strike ridge
Brandon, Vt.

Strongly dissected mountainous highland
Katahdin, Me.

MOUNTAIN FEATURES – Continued

Structurally controlled ridges
Tyrone, Pa.

Synclinal mountain
Tyrone, Pa.

Water gap
Cumberland, Md., W. Va.
Tyrone, Pa.

Wind gap
New Britain, Conn.

PLAINS FEATURES

Aggraded desert plain
Sonoma Range, Nev.

Alluvial plain
(*See* Valley Features)

Desert plain
Antelope Peak, Ariz.

Dissected lacustrine plain
Maumee, Ohio

Dissected plain
Mobile, Ala.

Dissected till plain
Effingham, Ill.

Flood plain
(*See* Valley Features)

Glaciated plain
(*See* Glaciation Features)

Lacustrine plain
Jordan Narrows, Utah

Mississippi alluvial plain
Philipp, Miss.

Plains remnant
Maumee, Ohio

PLATEAU FEATURES (buttes, mesas, outliers, etc.)

Apex of Ozark Plateau
Ironton, Mo.

Badlands (dissected plateau)
Sheep Mountain Table, S. Dak.

Butte
Bright Angel, Ariz.

Cuesta
Mammoth Cave, Ky.

Dip slope
Cumberland, Md., W. Va.

Dissected plateau
Sheep Mountain Table, S. Dak.

Dissected plateau of strong relief
Bright Angel, Ariz.

Erosional remnant
Thousand Springs, Idaho

Erosion surface
Chief Mountain, Mont.

Mesa
Juanita Arch, Colo.

Outlier
Sheep Mountain Table, S. Dak.

Plateau
Bright Angel, Ariz.

Point
Bright Angel, Ariz.

Summit erosion surface remnants
Mount Bonneville, Wyo.

Table
Sheep Mountain Table, S. Dak.

SOLUTION FEATURES

Blind valleys (valley sinks)
Mammoth Cave, Ky.

Disappearing stream
(*See* Water Features)

Karst topography
Mammoth Cave, Ky.
Oolitic, Ind.

Sinks
Mammoth Cave, Ky.

VALLEY FEATURES (basins, channels, drainage patterns, fans, meanders, terraces, etc.)

Abandoned channels
East Brownsville, Tex.
Thousand Springs, Idaho

Abandoned meander
Oolitic, Ind.

Alluvial fan
Ennis, Mont.

Alluvial fan, coalescing
Furnace Creek, Calif.

Alluvial fan, dissected
Furnace Creek, Calif.

Alluvial plain
Antelope Peak, Ariz.
Philipp, Miss.

Amphitheater
Bright Angel, Ariz.

Anticlinal valley
Altoona, Pa.

Arroyo
Antelope Peak, Ariz.

Bajada
Antelope Peak, Ariz.
Sonoma Range, Nev.

Barranca
Antelope Peak, Ariz.

Basin
Guadalupe Peak, Tex.
Sonoma Range, Nev.
Warm Springs, Ga.

Bed of drained shallow lake
Campti, La.

Bolson
Furnace Creek, Calif.
Guadalupe Peak, Tex.
Sonoma Range, Nev.

Canyon
Bright Angel, Ariz.
Juanita Arch, Colo.

Centrifugal drainage (originates on map)
Ironton, Mo.

Cove
Mount Desert, Me.

Dendritic drainage
Effingham, Ill.

Disappearing intermittent streams
Mammoth Cave, Ky.

Dissected terraces
Altoona, Pa.
Tyrone, Pa.

Entrenched meander
Oolitic, Ind.

Faultline valley
Bright Angel, Ariz.

Flats
Guadalupe Peak, Tex.

Flood plain
Menan Buttes, Idaho

Gorge
Bright Angel, Ariz.
Cumberland, Md., W. Va.

Gorge, postglacial
Catskill, N.Y.

Gully deeply eroded
Ship Rock, N. Mex.

Meander core
Oolitic, Ind.

Meander patterns
Philipp, Miss.

Meanders
East Brownsville, Tex.
Strasburg, Va.

Meander scars
Philipp, Miss.

Meandering stream in flood plain
Menan Buttes, Idaho

Narrows
Cumberland, Md., W. Va.
Jordan Narrows, Utah

Natural levee
Campti, La.
East Brownsville, Tex.

Nonintegrated drainage
Ashby, Neb.

Parallel drainage
Antelope Peak, Ariz.

Pediment
Antelope Peak, Ariz.

Playa
Furnace Creek, Calif.
Guadalupe Peak, Tex.
Sonoma Range, Nev.

Reverse drainage
Catskill, N.Y.

Rock terrace
Bright Angel, Ariz.

Salt basin
Guadalupe Peak, Tex.

VALLEY FEATURES (basins, channels, etc.) — Continued

Slip-off slope
Mammoth Cave, Ky.
Oolitic, Ind.

Stream piracy (capture)
Cumberland, Md., W. Va.
New Britain, Conn.

Strike valley
Brandon, Vt.
Catskill, N.Y.

Structurally controlled dissected terraces
Altoona, Pa.
Tyrone, Pa.

Structurally controlled drainage
Cumberland, Md., W. Va.

Structurally controlled valleys
Tyrone, Pa.

Synclinal valley
Strasburg, Va.
Tyrone, Pa.

Terrace, alluvial
Ennis, Mont.

Tonto Platform
Bright Angel, Ariz.

Trellis drainage
Cumberland, Md., W. Va.
Waldron, Ark.

Undercut slope
Mammoth Cave, Ky.
Oolitic, Ind.

V-shaped valley
Bright Angel, Ariz.

Wash or channel, sandy
Ship Rock, N. Mex.

Wide meander belts
Strasburg, Va.

VOLCANIC FEATURES

Ancient Mount Mazama
Crater Lake National Park and vicinity, Oreg.

Caldera
Crater Lake National Park and vicinity, Oreg.

Cinder cones
Menan Buttes, Idaho

Collapsed volcanic cone
Crater Lake National Park and vicinity, Oreg.

Crater
Crater Lake National Park and vicinity, Oreg.
Menan Buttes, Idaho

Dikes, radial
Ship Rock, N. Mex.

Dissected volcano
Crater Lake National Park and vicinity, Oreg.

Lava, recent
Menan Buttes, Idaho

Nuees ardentes deposits
Crater Lake National Park and vicinity, Oreg.

Parasitic cone
Crater Lake National Park and vicinity, Oreg.

Radial drainage on volcanic cone
Mount Rainier, Wash.

Rim of caldera
Crater Lake National Park and vicinity, Oreg.

Shield volcano
Crater Lake National Park and vicinity, Oreg.

Volcanic cone
Mount Rainier, Wash.

Volcanic cone, breached
Crater Lake National Park and vicinity, Oreg.

Volcanic cone within caldera
(Wizard Island)
Crater Lake National Park and vicinity, Oreg.

Volcanic necks and plugs
Ship Rock, N. Mex.

WATER FEATURES (lakes, streams, etc.)

Abandoned channels
(*See* Valley Features)

Abandoned river mouth
Fennville, Mich.

Artificial drainage
Jackson, Mich.

Bayou
(*See* Coastal Features)

WATER FEATURES (lakes, streams, etc.) — Continued

Braided stream
Ennis, Mont.

Cut-off meanders
Fennville, Mich.
Philipp, Miss.

Disappearing stream
Mammoth Cave, Ky.
Oolitic, Ind.

Finger lake
(*See* Glaciation Features)

High water table
Ashby, Neb.

Kettle with lake
(*See* Glaciation Features)

Lake in drowned tributary valley
Campti, La.

Lakes in sinks
(*See* Solution Features)

Mineral springs
Warm Springs, Ga.

Morainal lakes
(*See* Glaciation Features)

Oxbows
East Brownsville, Tex.

Oxbow lake
Campti, La.
East Brownsville, Tex.

Oxbow swamp
Philipp, Miss.

Ponds, glacial
(*See* Glaciation Features)

Poorly integrated drainage
(*See* Glaciation Features)

Rapids
Maumee, Ohio
Thousand Springs, Idaho

River with sand channel
(*See* Valley Features)

Slough
Campti, La.
Menan Buttes, Idaho

Springs
Crater Lake National Park and vicinity, Oreg.
Thousand Springs, Idaho

Stream piracy
(*See* Valley Features)

Superposed stream
Cumberland, Md., W. Va.
Ennis, Mont.
Warm Springs, Ga.

Water gap
(*See* Mountain Features)

WIND FEATURES

Blowout dune
Fennville, Mich.

Buried town (shifting sands)
Fennville, Mich.

Clay dune
East Brownsville, Tex.

Dune ridges, some transverse
Ashby, Neb.

Dunes and beach ridges
Jacksonville Beach, Fla.

Dunes and deflation hollows
Sheep Mountain Table, S. Dak.

Dune topography
Ashby, Neb.

Lakeshore dunes
Fennville, Mich.

Large-scale dune ridges
Ashby, Neb.

Sand bars and scrolls
(*See* Valley Features)

Sand dunes
Fennville, Mich.
Provincetown, Mass.

Sand hills
Ashby, Neb.

Map of Continental United States, showing locations of the 50 topographic maps

FIFTY TOPOGRAPHIC MAPS OF CONTINENTAL UNITED STATES

	State	Quadrangle		State	Quadrangle
1.	Alabama	Mobile	26.	Minnesota	Virginia
2.	Arizona	Antelope Peak	27.	Mississippi	Philipp
3.	Arizona	Bright Angel	28.	Missouri	Ironton
4.	Arkansas	Waldron	29.	Montana	Chief Mountain
5.	California	Furnace Creek	30.	Montana	Ennis
6.	California	Point Reyes	31.	Nebraska	Ashby
7.	Colorado	Juanita Arch	32.	Nevada	Sonoma Range
8.	Connecticut	New Britain	33.	New Hampshire	Monadnock
9.	Delaware	Little Creek	34.	New Mexico	Ship Rock
10.	District of Columbia, Va.	Washington West	35.	New York	Catskill
			36.	North Dakota	Pelican Lake
11.	Florida	Jacksonville Beach	37.	Ohio	Maumee
12.	Georgia	Warm Springs	38.	Oregon	Crater Lake National Park
13.	Idaho	Menan Buttes			
14.	Idaho	Thousand Springs	39.	Pennsylvania	Altoona
15.	Illinois	Effingham	40.	Pennsylvania	Tyrone
16.	Indiana	Oolitic	41.	Rhode Island	Kingston
17.	Kentucky	Mammoth Cave	42.	South Carolina	Mullins
18.	Louisiana	Campti	43.	South Dakota	Sheep Mountain Table
19.	Maine	Katahdin	44.	Texas	East Brownsville
20.	Maine	Mount Desert	45.	Texas	Guadalupe Peak
21.	Maryland, W. Va.	Cumberland	46.	Utah	Jordan Narrows
22.	Massachusetts	Lynn	47.	Vermont	Brandon
23.	Massachusetts	Provincetown	48.	Virginia	Strasburg
24.	Michigan	Fennville	49.	Washington	Mount Rainier
25.	Michigan	Jackson	50.	Wyoming	Mount Bonneville

1 ALABAMA
MOBILE QUADRANGLE 1940

location

Southwestern Corner of Alabama
Coastal Plain Province
East Gulf Coast Plain Section

physiographic features

* Abandoned Pleistocene shoreline (Pamlico Shoreline at foot of bluffs at about 30-foot level)

  Bay (Mobile Bay, Polecat Bay, etc.)

  Bayou (Alligator Bayou, etc.)

* Deltaic channels (Spanish River, Mobile River, etc.)

  Dissected plain (only eastern bluffs of the plain appear on this portion of the quadrangle, at Cottage Hill, Spring Hill, and Pine Grove)

* Distributary channels (Choctaw Pass, Pinto Pass)

* Drowned valley (Dog River and tributaries)

  Marine terrace (Pamlico Terrace, swampy, adjacent to Mobile Bay below the 30-foot level)

  Partially obstructed outlet (Dog River)

  Swamp above tidewater (Wragg Swamp)

  Tidal marsh or swamp (on Blakeley Island, etc.)

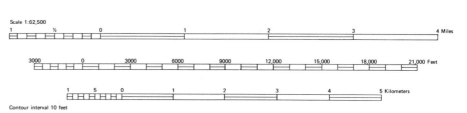

Scale 1:62,500

Contour interval 10 feet

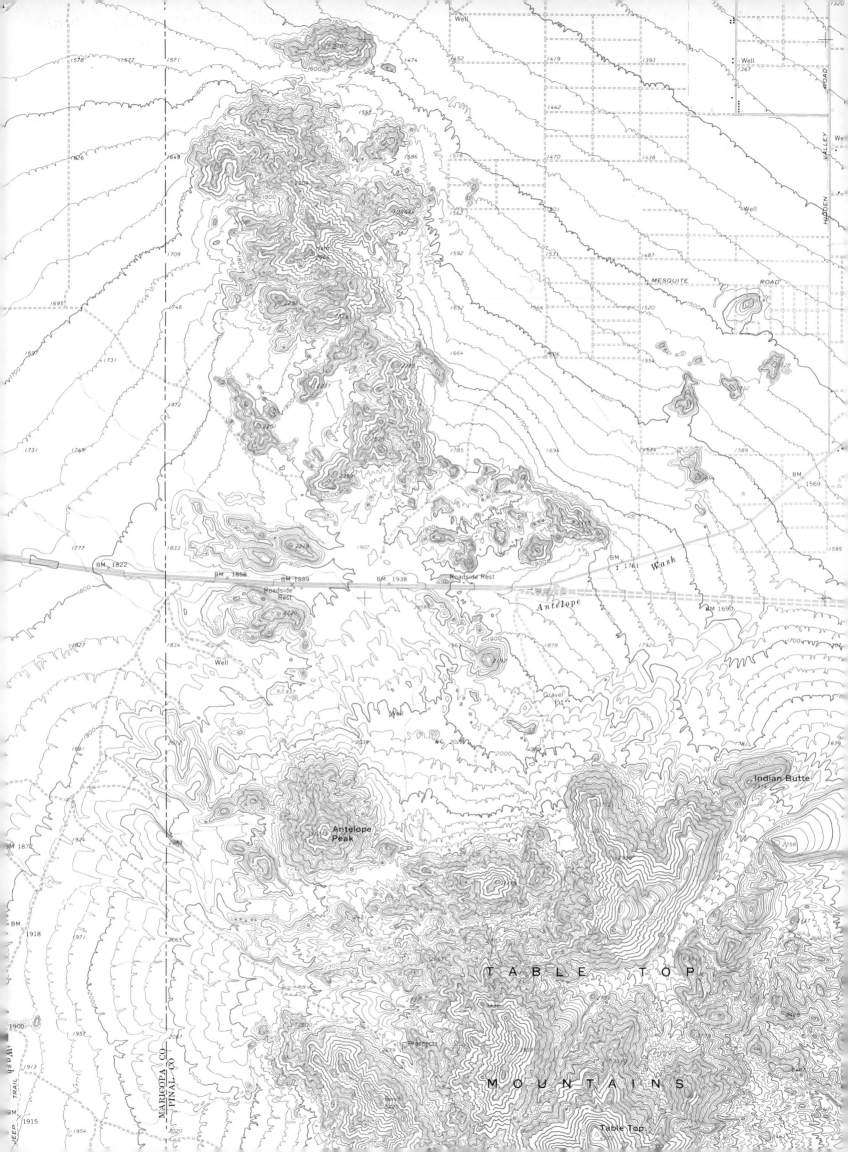

2

ARIZONA
ANTELOPE PEAK QUADRANGLE   1946

location

Southern Arizona
Basin and Range Province
Sonoran Desert Section

physiographic features

Alluvial fan (southwest corner of map)

* Alluvial plain (north part of map)

* Barranca (drains northeast area of Table Top Mountains from junctions of drains opposite elevation 2712; terminates alongside of Indian Butte)

* Desert plain (north portion of map)

* Erosional remnant (Antelope Peak)

* Inselbergs (north of BM 1931)

* Short range (Table Top Mountains)

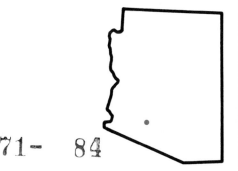

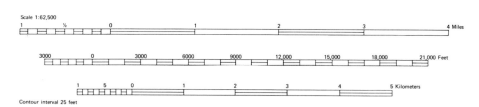

Contour interval 25 feet

3

ARIZONA
BRIGHT ANGEL QUADRANGLE   1903

location

Northwestern Arizona
Colorado Plateaus Province
Grand Canyon Section

physiographic features

* Butte (Dana Butte, Pattie Butte, etc.)
* Canyon (the Grand Canyon of the Colorado River; more than 4600 feet deep at BM 2436)
* Cliffs, accentuated by banded contouring (entire map)
* Dissected plateau of strong relief (entire map)
* Faultline valley (Bright Angel Canyon)
* Gorge (Granite Gorge, all of the Grand Canyon below Tonto Platform)
* Migrating divide (south rim of the Grand Canyon at Grandeur Point, Maricopa Point, etc.)
* Plateau (Coconino Plateau, southwest corner of map)
* Platform (Tonto Platform, lower rim of the Canyon at Plateau Point)
* Point (Grandeur Point, Maricopa Point, etc.)
  Rapid (Grapevine Rapids and many others without names)
* Rock terrace (below Cheops Pyramid)
* V-shaped valley (Granite Gorge, Bright Angel Creek, etc.)

note

The wide contour lines have absorbed others to indicate thickness of cliffs.

Scale 1:48,000

Contour interval 50 feet

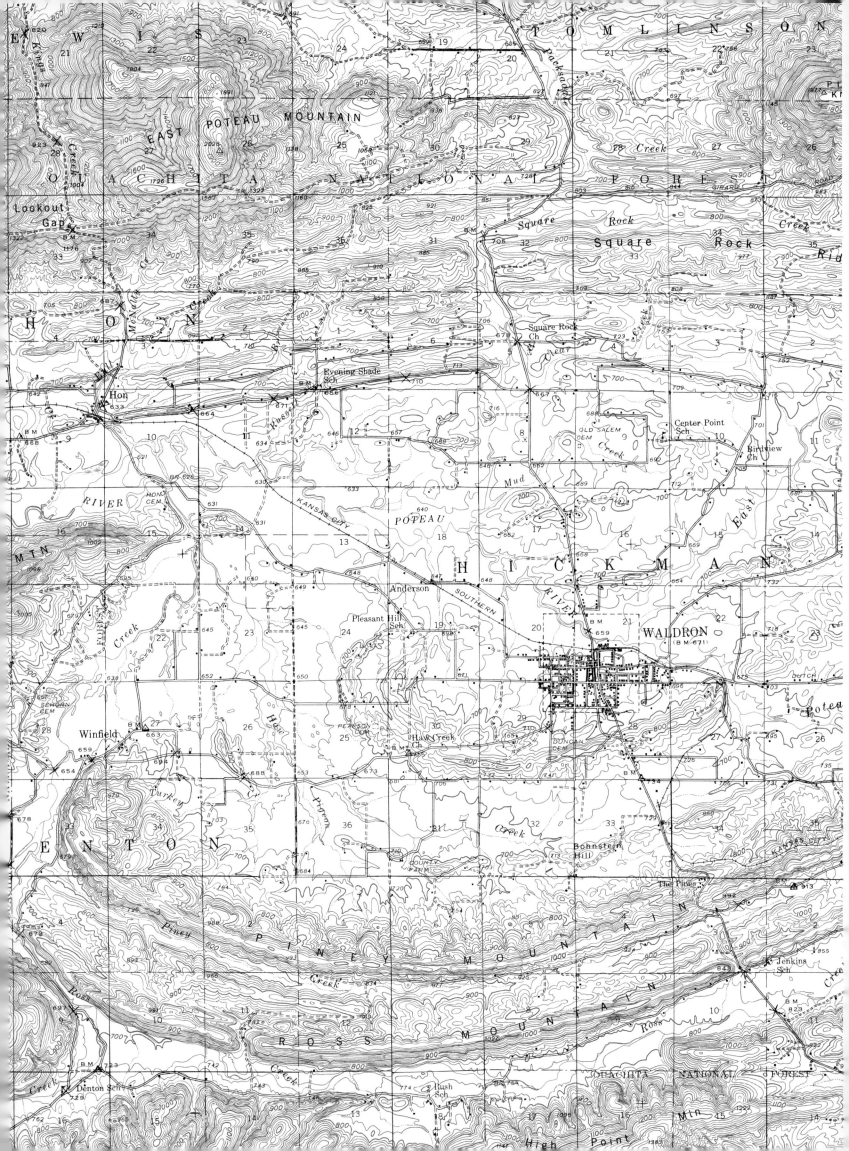

4 ARKANSAS
WALDRON QUADRANGLE 1937-1939

location

West-central Arkansas
Ouachita Province
Ouachita Mountains Section

physiographic features

* Folded mountain (Piney Mountain, etc.)

* Folds en echelon (series of parallel ridges beginning at Evening Shade School, extending northward through Square Rock Ridge, etc.)

* Hogback (Piney Mountain, Ross Mountain, etc.)

  Knob (Pilot Knob, northeast corner of map)

* Ridges formed of folded hard strata (Square Rock Ridge, etc.)

* Slip-off slope (point of ridge south of Hon Cemetery)

  Structurally controlled drainage (Ross Creek circling Ross Mountain, Packsaddle Creek, Square Rock Creek, etc.)

* Trellis drainage (Square Rock Creek, etc.)

  Undercut slope (south of Hon Cemetery near elevation 695)

  Water gap (Poteau River at southeast corner of Waldron)

  Wind gap (Lookout Gap)

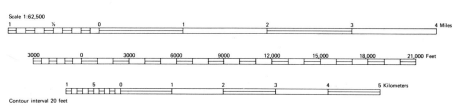

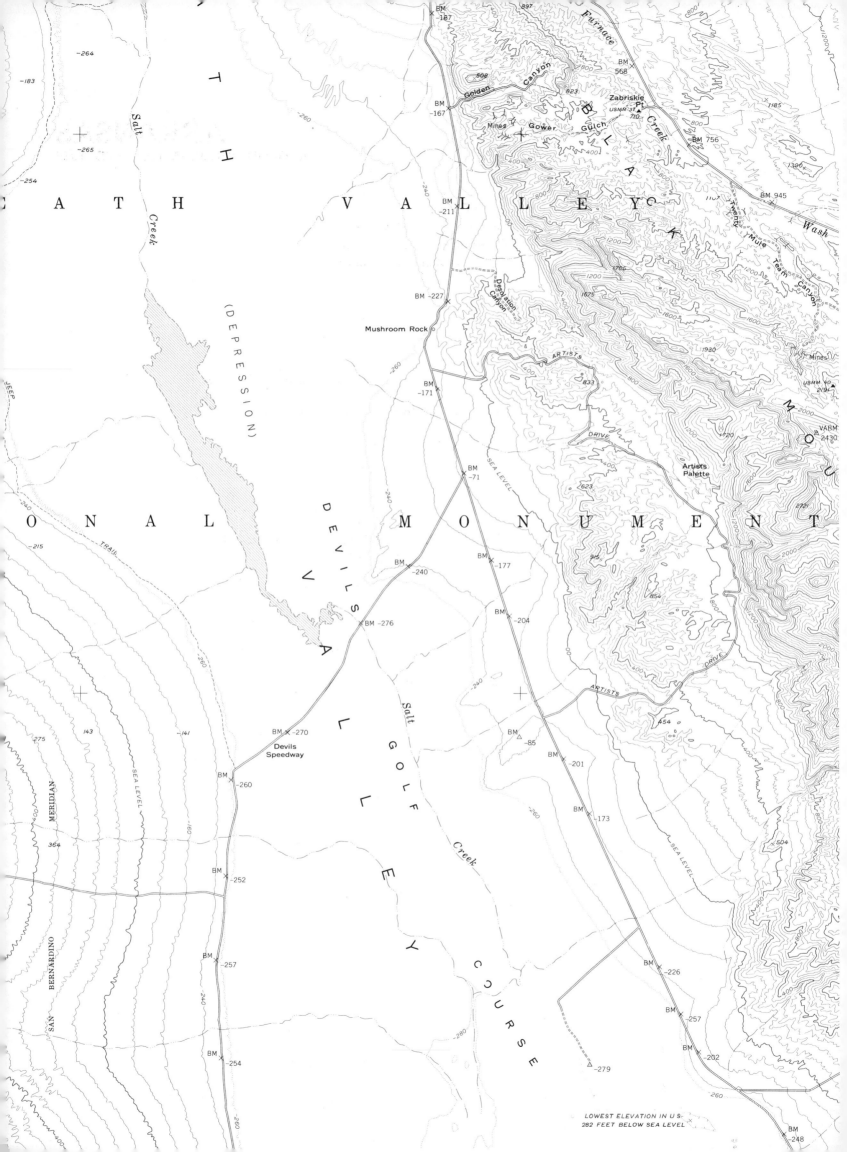

5

CALIFORNIA
FURNACE CREEK QUADRANGLE 1952

location

Southeastern California (in Death Valley National Monument)
Basin and Range Province
Great Basin Section

physiographic features

* Alluvial fan, dissected (southwest corner of map)

* Bajada (on both sides of Death Valley)

* Bolson (Death Valley)

  Dissected block mountain (Black Mountains, etc.)

  Dissected foothill, fan shaped (vicinity of Artists Drive)

* Faultline scarp (west face of Black Mountains, above Artists Drive)

* Isolated mountain range (Black Mountains, east side of Death Valley; the base of Panamint Range rises from the west side of the valley upward to the 11,649-foot Telescope Peak)

* Lowest elevation in the United States (minus 282 feet below sea level was determined in 1952)

  Mining area (vicinity of Twenty Mule Team Canyon)

* Playa (Dry Lake at the word "Depression")

  Sea level and below sea level contours to minus 280 contour (the below sea level contours are marked with a minus sign)

note

Prior to 1952 the lowest elevation had been accepted as −279 at nearby Badwater. The reason for the change in the lowest elevation is problematical, such as an unstable water table supporting the land surface.

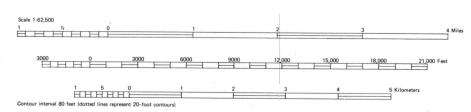

Scale 1:62,500

Contour interval 80 feet (dotted lines represent 20-foot contours)

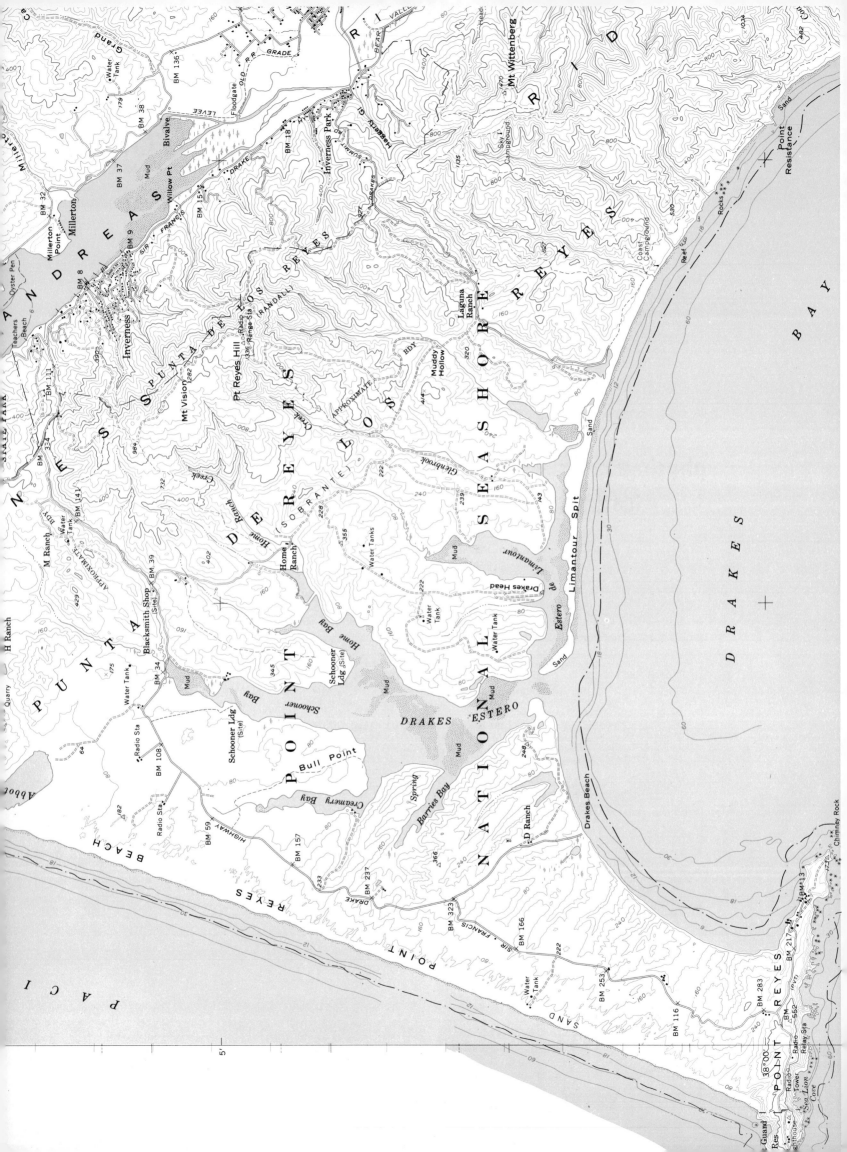

6 CALIFORNIA
POINT REYES QUADRANGLE 1950

location

Central Coast of California
Pacific Border Province
California Range Section

physiographic features

* Barrier beach, offset (at Drakes Estero eastward)
* Battered sea cliff (Point Reyes, Drakes Head, etc.)
  Bay (Drakes Bay; also Tomales Bay, the head of which is at Willow Point)
  Baymouth bar (at Drakes Estero)
  Block mountain (Inverness Ridge)
* Cape (Point Reyes)
  Delta (at head of Tomales Bay)
  Distributary stream on delta (at head of Tomales Bay)
  Drainage blocked by barrier beach (east of Drakes Head)
  Drowned valley (Tomales Bay)
* Estuary (Drakes Estero, Tomales Bay, etc.)
* Headland, truncated (Point Reyes)
* Lagoon (Abbotts Lagoon, Estero de Limantour, etc.)
  Marine terrace (along Drakes Bay, east part of map)
  Old Spanish land grant (Punta de Los Reyes)
  Parallel ridge and valley (numerous)
  San Andreas Rift (at Tomales Bay)
* Sand spit (from Drakes Estero eastward beyond Estero de Limantour; named Limantour Spit)
  Sea stacks (offshore from Point Reyes)
  Tidal flat (from Willow Point southward)
* Wave-cut cliff (Point Reyes, etc.)

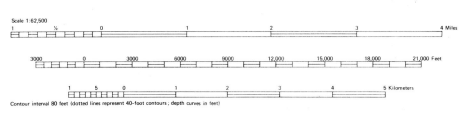

Contour interval 80 feet (dotted lines represent 40-foot contours; depth curves in feet)

27

7 COLORADO
JUANITA ARCH QUADRANGLE 1949

location

Western Border of Colorado
Colorado Plateaus Province
Canyon Lands Section

physiographic features

Canyon, V-shaped (lower Maverick Canyon, etc.)

* Canyon or gorge (Dolores River, spectacular as well as a fine example for study)

Cliff (*see* Escarpment)

Dry falls below hanging valley (notably on intermittent tributary streams west of Dolores River)

* Escarpment (at both rims of Dolores River, etc.)

* Esplanade (Flat Top Mesa)

* Mesa (Tenderfoot Mesa, etc.)

* Natural bridge (Juanita Arch)

Scale 1:24,000

Contour interval 20 feet

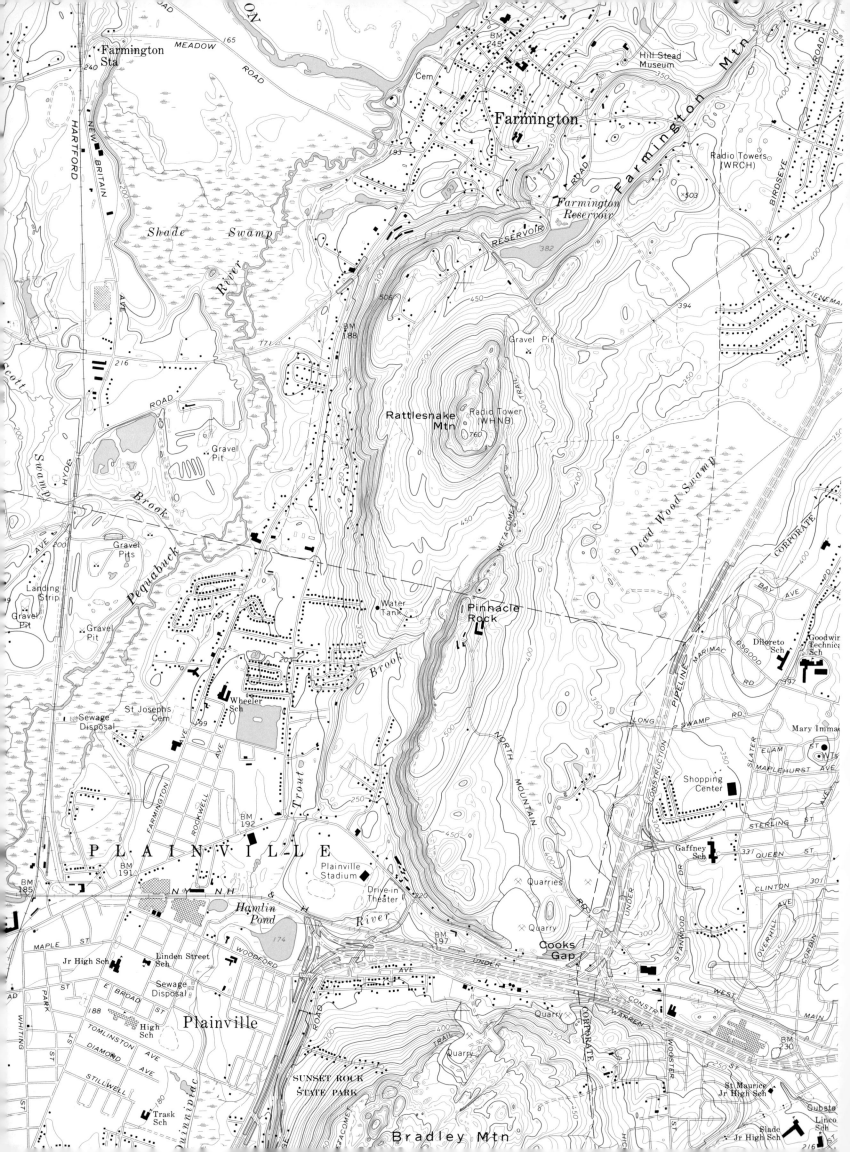

8 CONNECTICUT
NEW BRITAIN QUADRANGLE 1953

location

Central Connecticut
New England Province
New England Upland Section

physiographic features

Alluvial plain (northwest portion of map)

Artificial pond (near Wheeler School)

* Drainage reversal, glacial (Pequabuck River flowing north where Farmington River formerly flowed south)

* Faultblock mountain (Bradley Mountain, etc.)

Lava with irregular surface (Bradley Mountain, etc.)

* Outwash terrace (near Shade Swamp, etc.)

Pond (Hamlin Pond)

* Stream piracy (Quinnipiac River at Plainville, etc.)

Swamp (Dead Wood Swamp, Shade Swamp, etc.)

* Wind gap (Cooks Gap, ancient water gap of Ancestral Connecticut River)

note

The Farmington River is the wide stream at the north of the map flowing eastward toward the town of Farmington to its junction with the Pequabuck River, then northerly alongside the town.

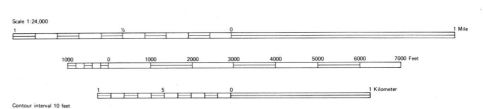

Scale 1:24,000

Contour interval 10 feet

9 DELAWARE
LITTLE CREEK QUADRANGLE 1946

location

Central Delaware
Coastal Plain Province
Embayed Section

physiographic features

Artificial drainage ditches (North South Canal and numerous smaller ditches in parallel patterns)

* Bay (Delaware Bay—shoreline outlines Little Bombay Hook, Kelly Island, etc.)

Cut-off bends (along Leipsic River)

Distributary channel (Old Womans Gut, Devers Gut, etc.)

Nearly abandoned distributary (Old Womans Gut)

* Pamlico Terrace (coastal terrace lying between the 6-foot and 8-foot levels)

* Silver Bluff Terrace (swampy, lying below the 6-foot level)

* Tidal drainage, dendritic (best examples are between Simons River and Old Womans Gut)

* Tidal meanders (throughout swampy areas)

* Tidal swamp or marsh (throughout lowland areas)

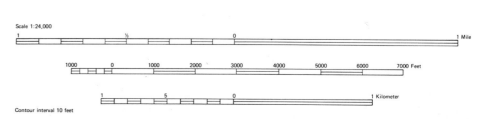

Scale 1:24,000

Contour interval 10 feet

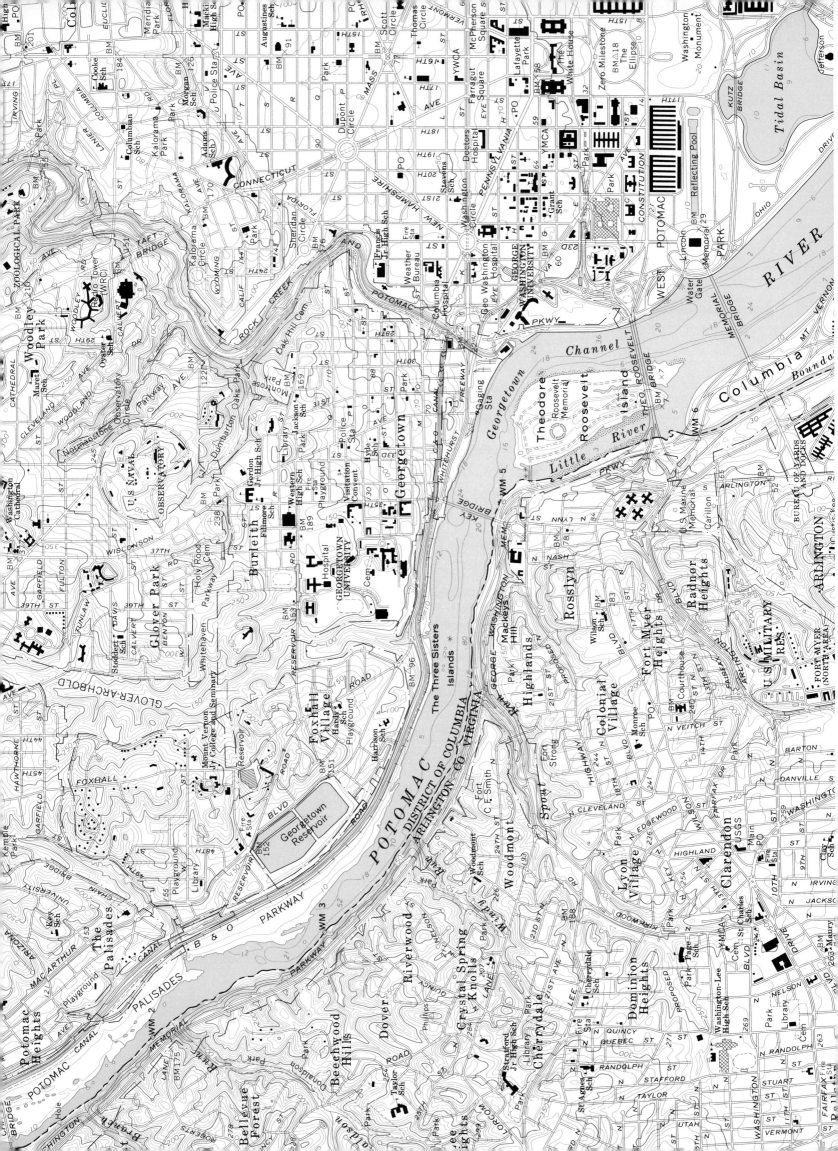

10 DISTRICT OF COLUMBIA
WASHINGTON WEST QUADRANGLE 1949

location

District of Columbia (also Maryland and Virginia)
Between Coastal Plain Province and Piedmont Province

physiographic features

Abandoned canal, historic (Chesapeake and Ohio Canal)

Artificial reservoir (Georgetown Reservoir)

Dissected depositional surface (Coastal Plain below Florida Avenue)

Dissected erosional surface (Piedmont above Florida Avenue)

* Drowned river (Potomac River below Key Bridge, also head of tide at Key Bridge)
* Entrenched meander (Rock Creek in National Zoological Park)

Entrenched stream (Potomac River above Key Bridge)

* Estuary (below Key Bridge)
* Faceted river bluffs (Palisades of Potomac River, south bank, above Key Bridge)
* Fall line (dividing line between Piedmont and Coastal Plain, along Florida Ave., etc.)
* Rocks awash (Three Sisters)

note

Key Bridge is named for Francis Scott Key

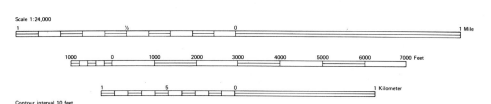

Scale 1:24,000

Contour interval 10 feet

11

FLORIDA
JACKSONVILLE BEACH QUADRANGLE 1948

location

Northern Coast of Florida
Coastal Plain Province
Floridian Section

physiographic features

* Ancient beach ridge (*see* note below)

* Barrier beach (present beaches and Silver Bluff Beach raised; *see* note)

* Beach (Manhattan Beach)

Depression (east of Greenfield Creek)

Dune and beach ridge (along Atlantic Ocean)

Drowned valley (Pablo Creek)

Indian mound (near Mayport Cemetery)

Prograded shore (Atlantic Ocean)

* Silver Bluff Beach, post-Pleistocene (below 6- to 8-foot level)

* Silver Bluff Shoreline (at elevation of 6 to 8 feet)

* Tidal swamp (along Intracoastal Waterway)

note

This is an ancient Coastal feature area, but Pamlico Beach, Coastal Bars, and Shoreline at elevation near 30-foot level are not on this portion of quadrangle. Neither is Pamlico Terrace which is above 30-foot level.

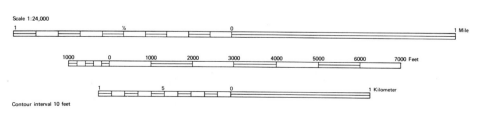

Scale 1:24,000

Contour interval 10 feet

12 GEORGIA
WARM SPRINGS QUADRANGLE 1934

location

West–central Georgia
Piedmont Province
Piedmont Upland Section

physiographic features

* Basin, eroded in soft rocks in a structural dome (The Cove)

  Dissected peneplain surface (most of map)

* Gorge (Flint River at Dripping Rocks, etc.)

  Resistant rock ridge, steeply dipping and folded (Pine Mountain, south part of map)

* Superposed stream (Flint River)

* Undercut slope (south bank of Flint River at foot of Rockhouse Mountain)

  Warm Springs (a well-known mineral springs and town of the same name)

* Water gap (on Flint River at Dripping Rocks, etc.)

  Wind gap (Dunn Gap, Patrick Gap, etc.)

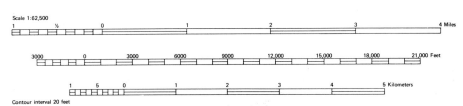

Scale 1:62,500

Contour interval 20 feet

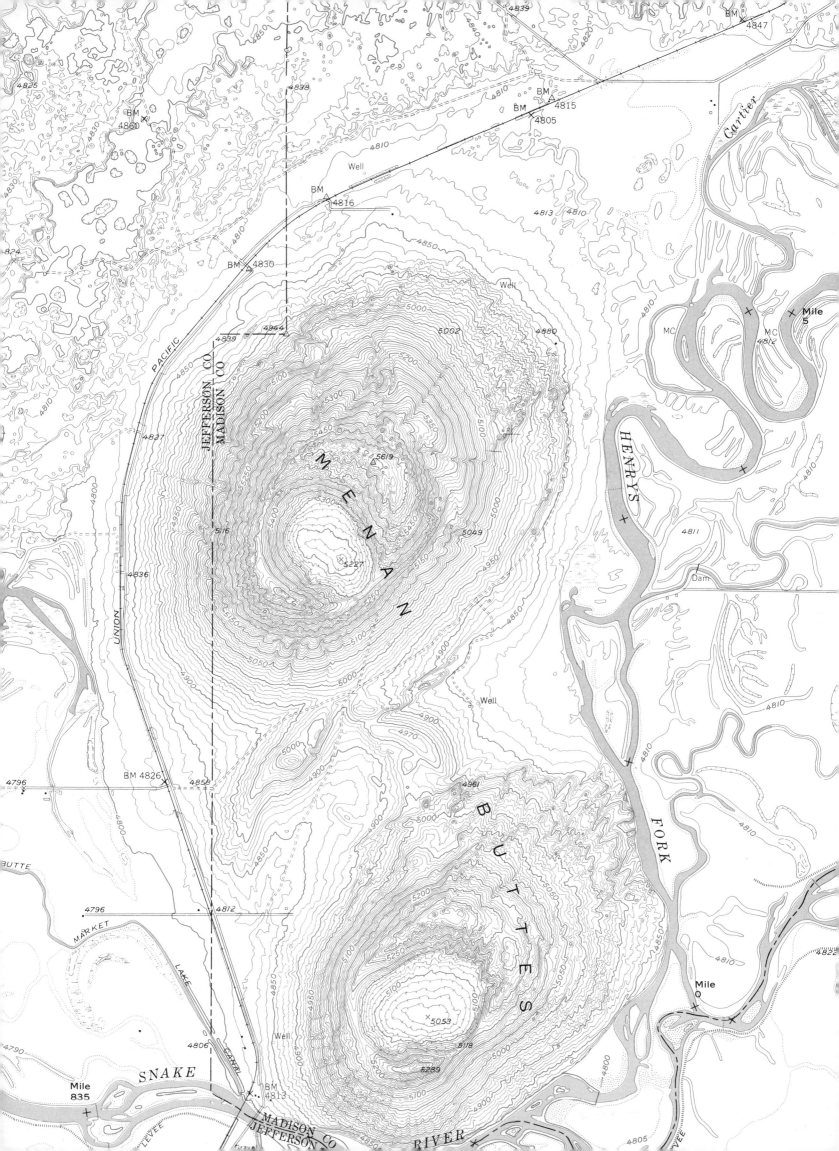

13 IDAHO
MENAN BUTTES QUADRANGLE   1951

location

Southeastern Idaho
Columbia Plateaus Province
Snake River Plain Section

physiographic features

Artificial levee (south part of map)

Check dam (east side of map)

* Cinder cones (Menan Buttes)

* Craters (Menan Buttes, extinct)

* Depression contour (in Menan Buttes craters and northwest area)

Flood plain (east side of map)

Irrigation canal (southwest corner of map)

* Lava flow, recent (northwest area of map)

* Meander scars (along Henrys Fork)

* Meandering stream (in flood plain)

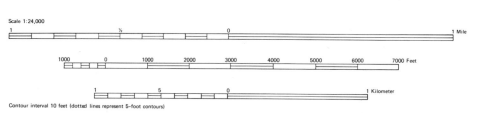

Scale 1:24,000

Contour interval 10 feet (dotted lines represent 5-foot contours)

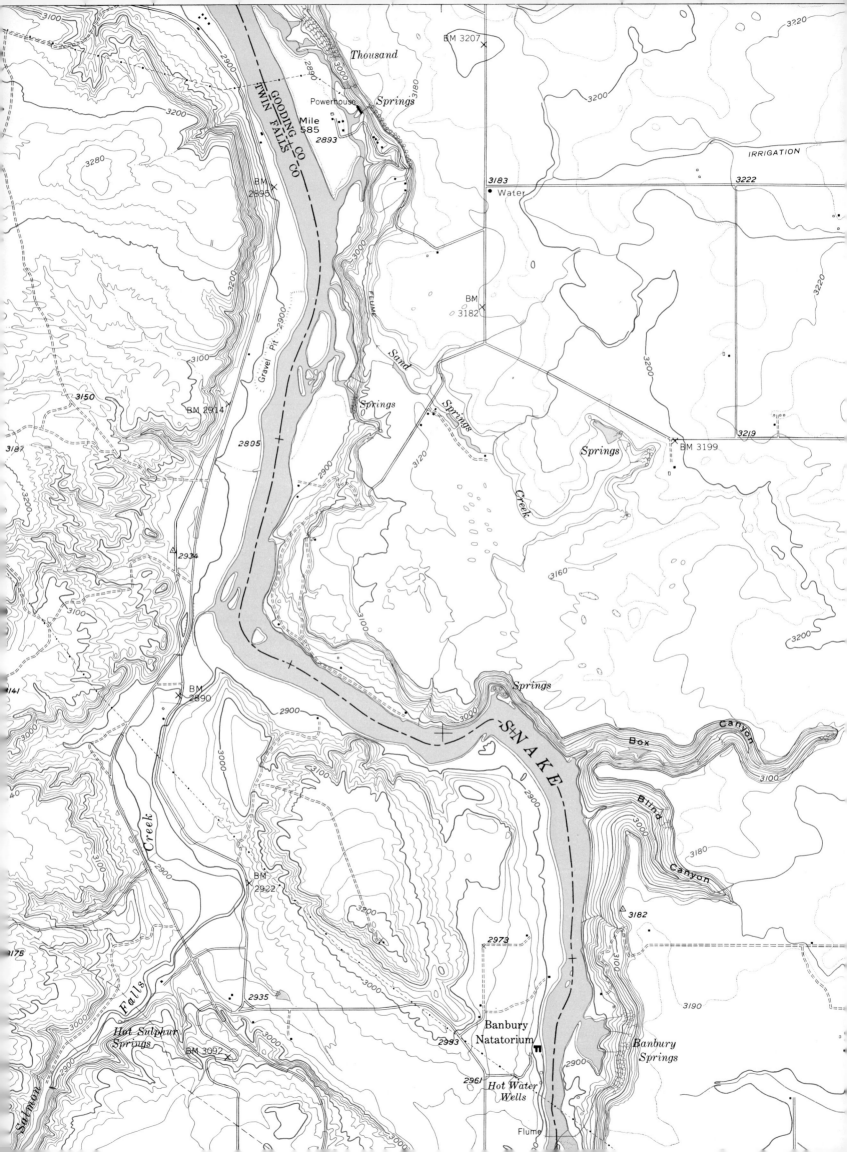

14

IDAHO
THOUSAND SPRINGS QUADRANGLE   1949

location

South-central Idaho
Columbia Plateaus Province
Snake River Plain Section

physiographic features

* Abandoned channel (former channel of Snake River from Snake River to Salmon Falls Creek)

* Canyon (Snake River at Thousand Springs, etc.; also Box Canyon and Blind Canyon)

* Erosional remnant (north of abandoned channel)

* Escarpments (along rim of Snake River at Thousand Springs, etc.; also at Box Canyon, etc.)

  Hot Springs (Hot Sulphur Springs)

  Plain (former wide valley filled with lava flows east of Snake River)

* Springs (various levels, right bank of Snake River; fed from underground streams following buried river channels and intercepted by north flowing tangent of Snake River)

* Springs captured for hydraulic power (Thousand Springs)

  Wells (Hot Water Wells)

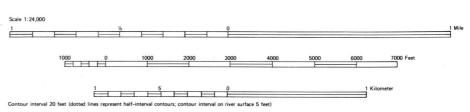

Scale 1:24,000

Contour interval 20 feet (dotted lines represent half-interval contours; contour interval on river surface 5 feet)

43

15 | ILLINOIS
EFFINGHAM QUADRANGLE 1950

location

South-central Illinois
Central Lowlands Province
Till Plains Section

physiographic features

* Dendritic drainage (especially the channel of Little Wabash River with symmetrical branches upstream from junction of Green Creek, Blue Point Creek, etc.)

Dissected till plain with mantle of loess deposits (over entire area)

End moraine (positioned by bedrock ridge just north of Shumway through Luthern Cemetery and Mound School)

* Gullied stream channels in flood plain (Henry Creek, Little Wabash River, etc.)

Integrated drainage (on end moraine)

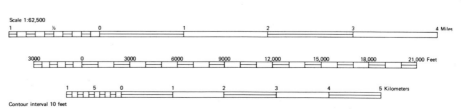

Contour interval 10 feet

16

INDIANA
OOLITIC QUADRANGLE 1935

Southern Indiana
Interior Low Plateaus Province
Possibly Western Section, Not Delimited

physiographic features

* Abandoned meander (near mouth of Crooked Creek)

* Disappearing streams (south portion of map)

* Dissected plateau (west portion of map)

* Entrenched meanders (White River)

* Karst topography (east portion of map)

  Limestone quarry (vicinity of Oolitic)

  Meander core (near mouth of Crooked Creek)

  Meander spur (Horseshoe Bend, etc.)

* Sinks, limestone (east portion of map)

* Slip-off slope (at Horseshoe Bend, etc.)

* Undercut slope (along White River opposite Horseshoe Bend, etc., and on Indian Creek north of Pleasant Hill Church)

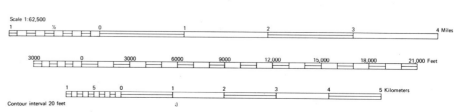

Scale 1:62,500
Contour interval 20 feet

17 KENTUCKY
MAMMOTH CAVE QUADRANGLE 1921–1922

location

Southwestern Kentucky
Interior Low Plateaus Province
Highland Rim Section

physiographic features

* Blind valley (Cedar Spring Valley, Woodside Hollow)
* Cave (Mammoth Cave, Colossal Cave, New Entrance to Mammoth Cave)

  Cuesta (Dripping Spring Cuesta from Little Hope School eastward to just beyond Brownsville Pike)
* Disappearing stream (Gardner Creek, Double Sink)
* Dissected plateau (most of north portion of map)
* Entrenched meander (best example at Turnhole Bend)
* Entrenched stream (Green River)
* Karst topography (two types: the upland dissected plateau area, northern part of map, is featured with large sinks; the southern lowland area, near base level of subterranean streams, has an abundance of small sinks, many with ponds)

  Knob (Goblers Knob)
* Sink (Hunts Sink, Cedar Sink, Double Sink—all in plateau area; a multitude of small sinks in the lowland southern area)

  Slip-off slope (above McCoy Landing, at Turnhole Bend, and opposite YMCA Camp)

  Spring (Dripping Spring)

  Undercut slope (Just above McCoy Landing, at Nappers Rollover)

Scale 1:62,500

Contour interval 20 feet

18

LOUISIANA
CAMPTI QUADRANGLE 1944

location

Northern Louisiana
Coastal Plain Province
West Gulf Coastal Plain Section

physiographic features

* Abandoned channel (north end of Crane River Lake)

Artificial levee (at Bethel School)

* Bayou (Bayou Pierre, etc.)

* Bed of drained shallow lake (Old Spanish Lowlands drained by Johnson Chute into Bayou Pierre)

Cut-bank (Bayou Pierre)

Dissected upland (east end of Sabine Uplift, south portion of map)

* Flood plain (Red River)

Flood plain swamp (Campti Brake, etc.)

Great Raft, a jam formed of timber naturally thrown into the stream by the caving of banks, creating a huge lake in a nearby drowned tributary of Red River called Black Lake (reported east and below Campti by early Spanish and French explorers, Black Lake is just east of this portion of the quadrangle)

Marsh or swamp (Campti Break, etc.)

Meandering stream (Red River)

* Oxbow lake (Old River)

River shore bars (above and below mouth of Bayou Pierre)

* Slough (Bayou Pierre, etc.)

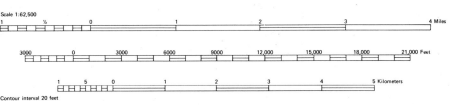

19 MAINE
KATAHDIN QUADRANGLE   1949

location

Northern Maine
New England Province
White Mountain Section

physiographic features

Arete (Knife Edge, near South Peak)

* Biscuit-board topography (grouping of North Basin, Great Basin, and South Basin)

Cirque (North Basin)

* Col (head of Great Basin and north of Saddle Spring)

* Cyclopean stairs (through and below North Basin)

Hanging valley (North Basin, etc.)

Kames (vicinity of Cranberry Pond, etc.)

Marsh or swamp (vicinity of Togue Ponds)

Meanders (Abol Deadwater)

* Monadnock (Mount Katahdin, elevation 5267, said to be the first point on which the morning sun shines on the Continental United States)

Ponds in kettles (Rocky Pond, etc.)

* Radial drainage (Mount Katahdin)

* Strongly dissected mountainous highland (north half of map)

Tarn (Davis Pond, Chimney Pond, etc.)

Waterfalls or rapids (Abol Falls, Pockwockamus Falls on Penobscot River)

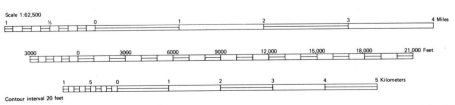

Scale 1:62,500

Contour interval 20 feet

20

MAINE
MOUNT DESERT QUADRANGLE 1942

location

Northern Coast of Maine
New England Province
Seaboard Lowland Section

physiographic features

   Bay (Eastern Bay, Western Bay)

   Cliff (Beach Cliff, etc.)

* Cove (Valley Cove, Goose Cove, etc.)

* Drowned coast line (entire map)

   Finger lakes (Echo Lake, etc.)

* Fiord (Somes Sound)

* Lakes and ponds in glacially scoured bedrock basin (Great Pond, Echo Lake, etc.)

   Marsh (over most of north end of Mount Desert Island)

* Mountains and island modified by continental glaciation (Bernard Mountain, Mount Desert Island, etc.)

   Narrows (Mount Desert Narrows, Bartlett Narrows, etc.)

* Rock sculpture controlled by fractures (at Seal Cove Pond, etc.)

   Tidal marsh (Jones Marsh, etc.)

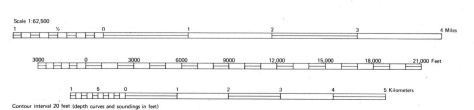

Scale 1:62,500

Contour interval 20 feet (depth curves and soundings in feet)

55

21 MARYLAND
CUMBERLAND QUADRANGLE 1944

location

Western Maryland (also minor portions of Pennsylvania and West Virginia)
Valley and Ridge Province
Middle Section

physiographic features

* Abandoned canal, historic (southeast corner of map)
* Allegheny Front (west of and parallel to Haystack Mountain)
* Anticlinal ridge (Haystack Mountain–Wills Mountain)
* Gorge (The Narrows)
- Hogback (Knobly Mountain, Shriver Ridge, etc.)
- Stream piracy (lower course of Braddock Run diverted by tributary of Wills Creek)
* Structurally controlled drainage (North Branch Potomac River)
* Superposed stream (Wills Creek at The Narrows)
* Trellis drainage (at Eckhart Junction, etc.)
* Water gap (The Narrows and at the Courthouse)
- Wind gap (Haystack Mountain at Braddock Road)

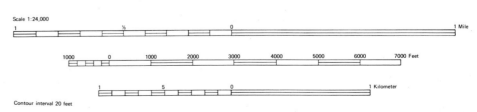

Scale 1:24,000

Contour interval 20 feet

22

MASSACHUSETTS
LYNN QUADRANGLE   1956

location

Massachusetts Coast East of Boston
New England Province
Seaboard Lowland Section

physiographic features

Artificial land (north of Saugus River and Pines River confluence)

* Bayhead bar (Pond Beach)

* Baymouth bar (Revere Beach)

* Tidal marsh or swamp (Pines River area up to long meander in Saugus River)

* Tombolo, complex (at Little Nahant and at the large area of Nahant south of Little Nahant Beach)

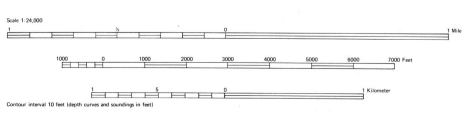

Scale 1:24,000

Contour interval 10 feet (depth curves and soundings in feet)

23 MASSACHUSETTS
PROVINCETOWN QUADRANGLE 1941

location

Southeastern Massachusetts
New England Province
Seaboard Lowland Section

physiographic features

Beach ridge (North coastline of Cape Cod)

* Cape (Cape Cod)

* Compound recurved spit (Cape Cod, including Long Point)

Cove (Herring Cove)

* Harbor (Provincetown Harbor)

* Lagoon (Hatches Harbor)

* Point (Long Point, Race Point)

Ponds in depressions (Clapps Pond, Clapps Round Pond, etc.)

Sand dunes (on most of map)

* Spit (Long Point, Race Point)

Tidal swamp (north of Wood End Lighthouse)

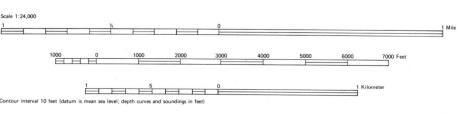

Contour interval 10 feet (datum is mean sea level; depth curves and soundings in feet)

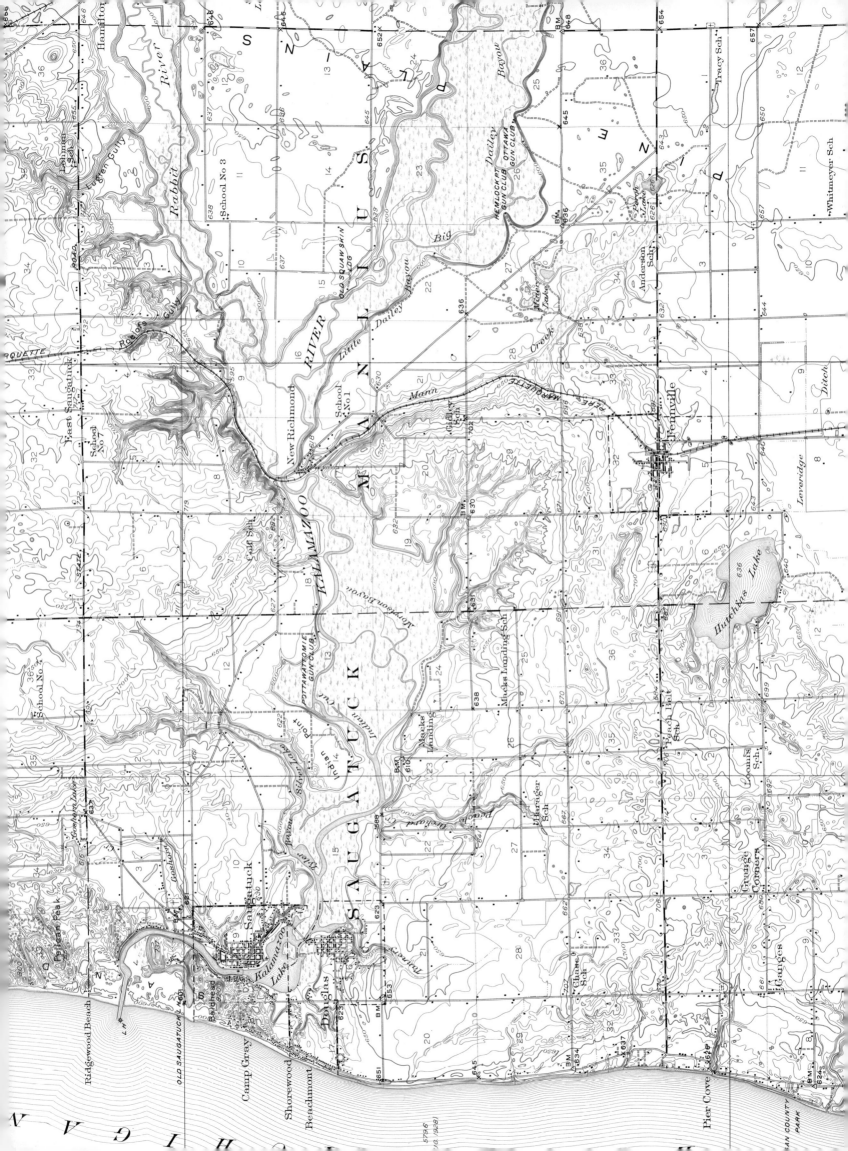

24 MICHIGAN
FENNVILLE QUADRANGLE   1928

location

Southwestern Michigan
Central Lowland Province
Eastern Lake Section

physiographic features

* Abandoned river mouth (at Old Saugatuck Lighthouse, etc.)
* Blowout dunes (vicinity of Pelican Peak)

  Buried town (north side of Kalamazoo River, former lumber town of Singapore, flourished 1834-1890 until buried by shifting sand)
* Cut-off meanders in Kalamazoo River (Morrison Bayou, etc.)

  Drainage canals (below Hutchins Lake, also Leveridge Ditch)

  Entrenched tributary (Roelofs Gully, etc.)
* Intermorainal lowland (east portion of map, south of Rabbit River)

  Knobs and kettles (vicinity of Hutchins Lake)

  Lake (Lake Michigan, one of the Great Lakes)
* Lake border moraine (west of Kalamazoo Lake, etc.)

  Lake in kettle (Perch Lake)
* Lakeshore dunes (Baldhead, 820 feet high, 240 feet above Lake Michigan)

  Morainic topography (west of Kalamazoo Lake, etc.)

  Pitted outwash plain (south of Rabbit River)
* Raised beach ridge (from Douglas southward)

  Raised spit and hook (at Douglas)

  River terrace (north bank of Rabbit River)
* Sand dunes (modern active dunes at lakeshore from Saugatuck north along Lake Michigan)

  Swamp (along Kalamazoo River flood plain, etc.)

  Valparaiso Moraine (east portion of map)
* Wave-cut cliffs (along Lake Michigan south from Beachmont)

Scale 1:62,500

Contour interval 10 feet (contour interval in sand dunes area 20 feet)

25

MICHIGAN
JACKSON QUADRANGLE 1939

location

Southern Michigan
Central Lowland Province
Eastern Lake Section

physiographic features

* Abandoned glacial channels (northwest part of map)

* Artificial drainage (incorporating poorly integrated Grand River from Vandercook Lake through Sharpes Lake, then through Sections 13 and 14)

* Esker (Blue Ridge near Skiff Lake to vicinity of Greens Lake)

* Kalamazoo Moraine (southwest portion of map)

* Kames (throughout Kalamazoo Moraine)

* Knobs and kettles (throughout Kalamazoo Moraine)

* Lake in kettle (Mud Lake, etc.)

* Morainic topography, terminal (Kalamazoo Moraine)

* Ponds in kettles (Kalamazoo Moraine)

* Poorly integrated drainage with many lakes (entire map)

Swamp or marsh (over large portion of map)

Till plain (north part of map)

City limits of Jackson, for which quadrangle is named, appear on north edge of map (including Ella Sharp Park)

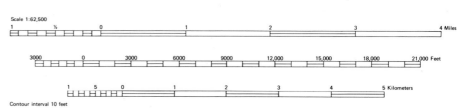

Scale 1:62,500

Contour interval 10 feet

26

MINNESOTA
VIRGINIA QUADRANGLE 1951

location

Northeastern Minnesota
Superior Upland Province

physiographic features

   Glacial drift (southwest corner of map)

* Glacially rounded hills (north portion of map)

   Kettles in outwash plain (southwest corner of map)

* Large open pit iron mine (Missabe Mountain Mine)

* Laurentian Divide (south portion of map)

   Linear ridge on crystalline rock (Laurentian Divide)

   Mine dumps and tailings ponds

   Monadnock (Lookout Mountain)

   Swamp or bog (at elevation 1536, east of Lookout Mountain and south of a water gap)

   Water gap (south of Lookout Mountain)

* Wind gap (on Laurentian Divide)

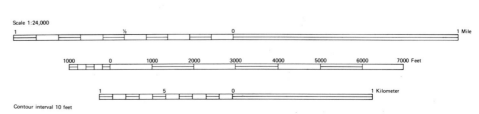

Scale 1:24,000

Contour interval 10 feet

27

MISSISSIPPI
PHILIPP QUADRANGLE 1932

location

Northwestern Mississippi
Coastal Plain Province
Mississippi Alluvial Plain Section

physiographic features

   Abandoned channel (Hardtime Bayou, etc.)

   Alluvial fans (at base of river bluff)

* Alluvial plain (most of map)

* Ancient river bluff, partially dissected (east side of map)

* Bayou (Tippo Bayou, etc.)

* Cut-off meander (several on Tippo Bayou, including Long Brake)

   Dissected loess upland (above river bluff)

   Flood plain swamp (large area on east part of map)

* Marsh or swamp (Hubbard Brake and the large flood plain swamp)

* Meander patterns (large: along Tallahatchie River; small: around Blue Lake, Sherrill Lake, etc.)

   Meander scars (along Tallahatchie River and Tippo Bayou)

* Meandering stream on alluvial plain (Tippo Bayou)

* Mississippi alluvial plain (known as "The Delta")

* Oxbow lakes (Hampton Lake, Yonkapin Lake, etc.)

   Oxbow swamp (at Blue Lake)

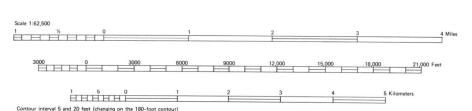

Scale 1:62,500

Contour interval 5 and 20 feet (changing on the 180-foot contour)

28

MISSOURI
IRONTON QUADRANGLE 1936-1937 and 1945

location

East-central Missouri, known as the "Iron Mountain Country"
Ozark Plateaus Province
Springfield-Salem Plateaus Section

physiographic features

* Apex of the Ozark Plateau (entire map)

* Centrifugal drainage (with all drains on this quadrangle originating within its boundaries)

* Dissected upland (entire map)

* Knob (Pilot Knob)

  Mine (south end of Iron Mountain and on Pilot Knob)

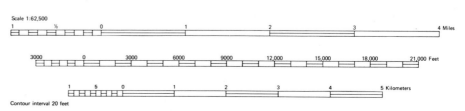

Scale 1:62,500
Contour interval 20 feet

29

MONTANA
CHIEF MOUNTAIN QUADRANGLE
MAPPED 1900–1902, REVISED 1938

location

Northwestern Montana (in Glacier National Park)
Rocky Mountain System Province
Northern Rocky Mountain Section

physiographic features

* Alpine topography (nearly all features made by Alpine glaciation are represented on this map)
* Arete (The Garden Wall, Ptarmigan Wall, etc.)
* Cirque (at Grinnell Glacier, Sue Lake, etc.)
* Cirque lake (Sue Lake, Iceberg Lake, etc.)
* Col (Swiftcurrent Pass)
* Cyclopean stairs (Ipasha Glacier to Mokowanis Lake)
* Finger lake (Glenns Lake, portion of Lake McDonald, etc.)
* Glacial trough (Mokowanis River)
* Glacier (Grinnell Glacier, etc.)
* Hanging valley (Hidden Lake, Elizabeth Lake, etc.)
* Klippe (Chief Mountain—a giant landmark and famous erosional remnant)
* Matterhorn (Going-To-The-Sun Mountain)
* Pass (Logan Pass, Swiftcurrent Pass, etc.)
* U-shaped valley (Belly River)

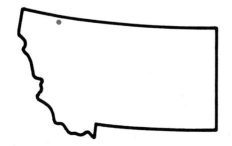

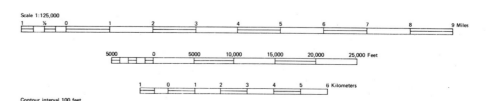

Contour interval 100 feet

30 | MONTANA
ENNIS QUADRANGLE 1949

location Southwestern Montana (near Yellowstone National Park)
Northern Rocky Mountains Province

physiographic features

* Alluvial fan (Cedar Creek Alluvial Fan, unusually large, covering more than 20 square miles and very symmetrical; formed from sand and gravel deposits below where Cedar Creek issues through a water gap)
* Braided stream (Madison River)
  Coalescing alluvial fans (south part of map, etc.)
* Flood plain (Madison River)
  Matterhorn (Fan Mountain)
* Radial drainage (on Cedar Creek Alluvial Fan)
* Terrace, alluvial (southwest of Shelhamer Ranch)
* Water gap (Cedar Creek above Lawton Ranch)

note The Madison River joins the Jefferson River and Gallatin River at Three Rivers, some 150 miles north of this area. It becomes the Missouri River at the triple confluence.

Scale 1:62,500

Contour interval 40 feet

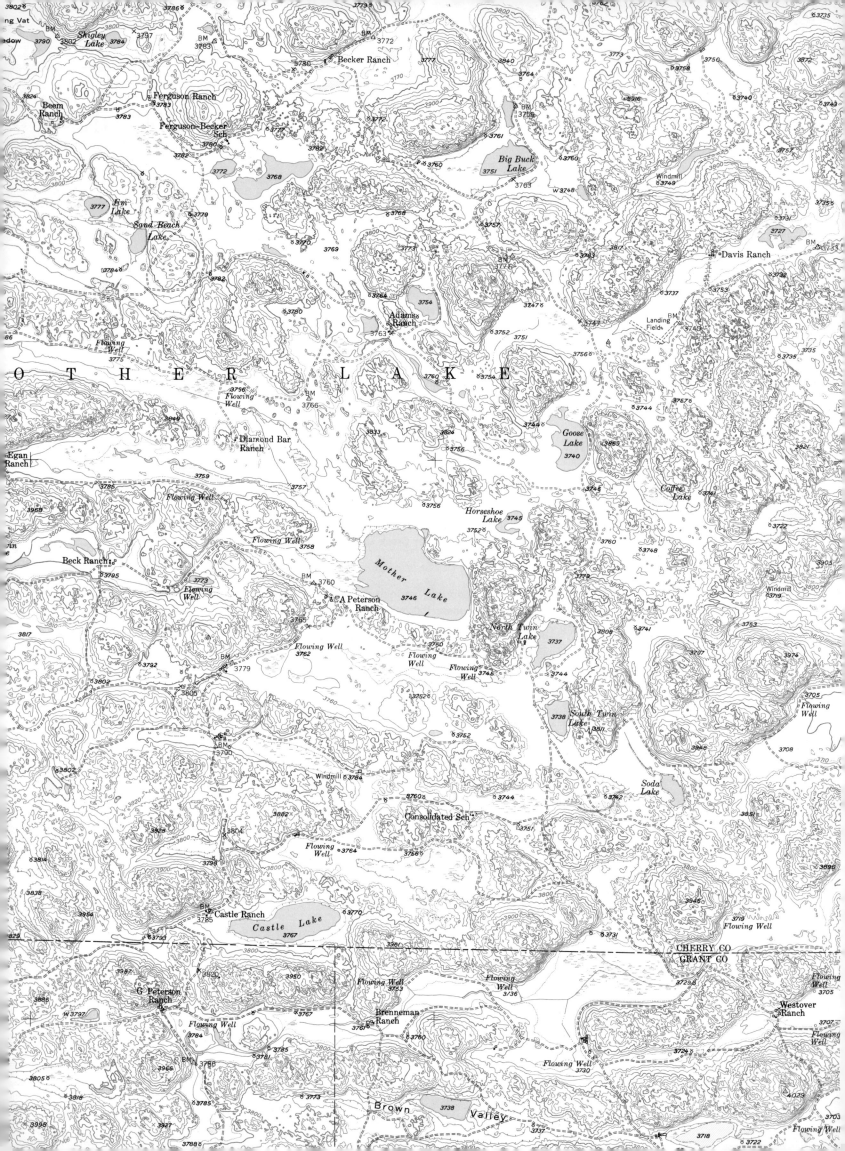

31

NEBRASKA
ASHBY QUADRANGLE 1948

location

West-central Nebraska
Great Plains Province
High Plains Section

physiographic features

Artificial drainage canal (through Sections 1, 2, 3, 4, 5, and 6, south of the Cherry-Grant county line)

Concave slope, exemplary of lee side (northwest part of Section 2 at county line, etc.)

Depression with lake (Castle Lake, etc.)

* Dune topography (entire map)

* High water table with flowing wells, lakes, and marshes (entire map)

* Large dune ridges, stabilized (some transverse and modified by secondary wind erosion following stabilization; numerous minor blowouts)

* Nonintegrated drainage (entire map)

* Sand hills of Nebraska

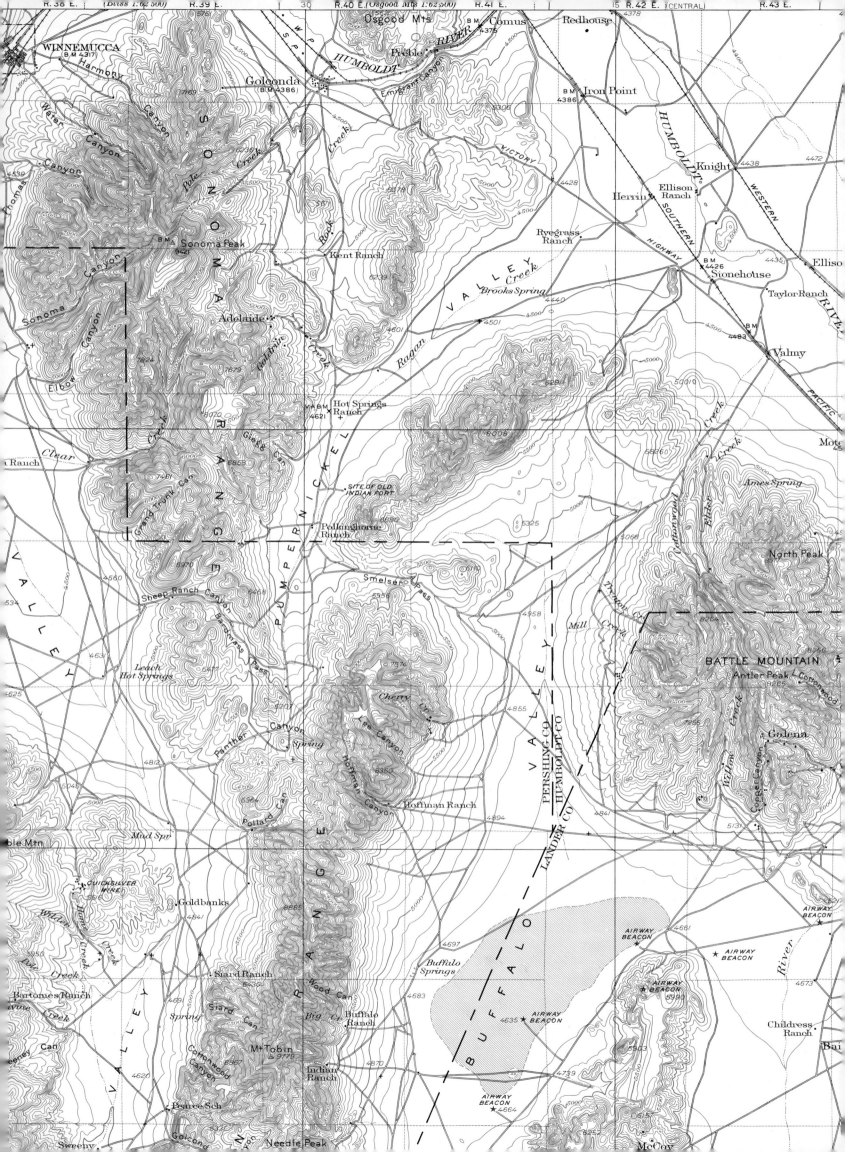

32

NEVADA
SONOMA RANGE QUADRANGLE   1932

location

Northern Nevada
Basin and Range Province
Great Basin Section

physiographic features

* Aggraded desert plain (Buffalo Valley, etc.)

  Alluvial fan (below Willow Creek which drains the north part of Battle Mountain)

* Bajadas (throughout map)

* Basin (Buffalo Valley)

* Bolson (Buffalo Valley)

* Dissected block mountain (Tobin Range, etc.)

  Fault scarp (along west base of Mount Tobin)

* Flat (at Buffalo Valley)

  Gap or pass (Bardmass Pass)

  Hot springs (Leach Hot Springs)

* Intermittent lake (adjacent to Buffalo Springs)

* Isolated range (Sonoma Range, etc., separated by aggraded desert plains)

* Playa (in Buffalo Valley)

  Springs (numerous for this semiarid country)

  Volcanic cone (at Airway Beacon, east of Buffalo Valley)

  Water gap (Emigrant Canyon on west-flowing Humbolt River)

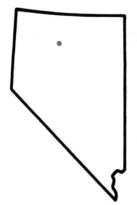

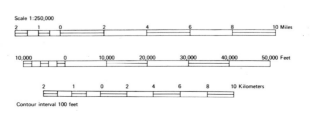

33

NEW HAMPSHIRE
MONADNOCK QUADRANGLE 1949

location

Southern New Hampshire
New England Province
New England Section

physiographic features

Bedrock hills in crystalline rocks abraded by continental glaciation (entire map)

* Concave slope (Monadnock Mountain)
* Drainage deranged by continental glaciation (vicinity of Stanford Pond, etc.)
* Drumlins (southeast part of map)
* Glacial drift (east of Monadnock Mountain)
* Hanging valley (Stone Pond, etc.)

  Kame terrace (east of Frost Brook, etc.)
* Kettle holes (Gibson Pond, etc.)
* Lakes or ponds in kettle holes (Dark Pond, etc.)
* Monadnock Mountain (feature from which the physiographic term Monadnock was derived)
* Morainal lake or pond (Thorndike Pond, etc.)
* Mountain peak, isolated (Monadnock Mountain)

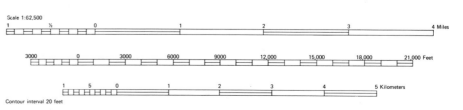

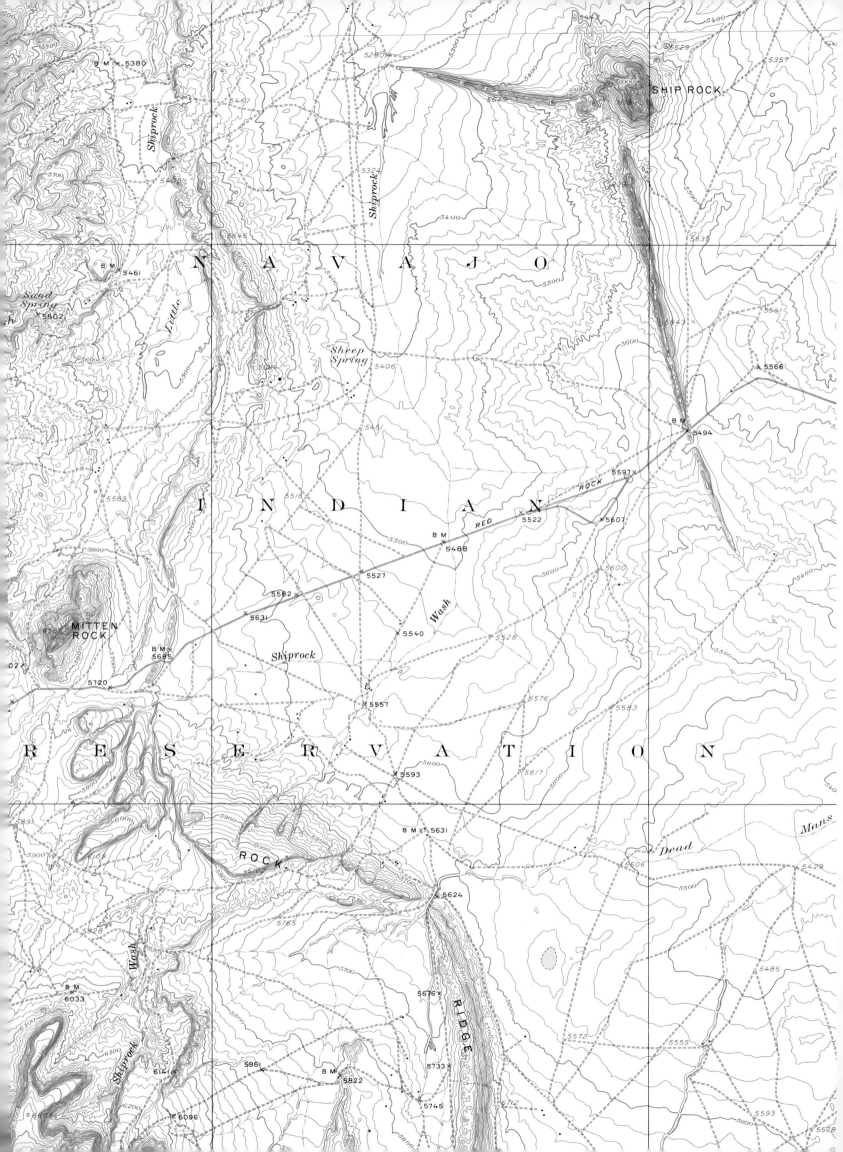

34

NEW MEXICO
SHIP ROCK QUADRANGLE 1934

location

Northwestern Corner of New Mexico
Colorado Plateaus Province
Navajo Section

physiographic features

* Dike, radial (south from Ship Rock through BM 5494; also west from Ship Rock to wash near elevation 5280)

* Dip slope (on west portion of Rock Ridge, heading at elevation 6102; also heading at nearby elevation 6109)

  Flatirons (along Rock Ridge beginning 1 mile east at elevation 6102)

* Gully, deeply eroded (Little Ship Rock Wash downstream from Mitten Rock; also Ship Rock Wash above BM 5685)

  Hogback (southern portion of Rock Ridge north of BM 5937)

* Volcanic neck (Ship Rock, Mitten Rock)

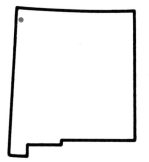

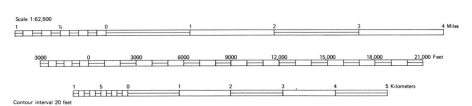

Scale 1:62,500

Contour interval 20 feet

35

NEW YORK
CATSKILL QUADRANGLE 1941

location

Southern New York
Valley and Ridge Province
Hudson River Section

physiographic features

* Bedrock knob (Blue Hill, Mount Merino)

  Cliffs (west side of map)

  Delta (at Saugerties)

  Disappearing stream (north of VanLuven Lake)

* Dissected glaciated plateau bounded on east by escarpment (Vedder Hill through Timmerman Hill to Mount Airy)

* Drowned river (Hudson River)

  Drumlins (Cross Hill, Roundtop, etc.)

* Elongated folds and ice-carved strike ridges (west portion of map)

  Glacial lake bed (Kiskatom Flats)

  Limestone quarry (vicinity of Alsen)

  Meandering stream (Roeliff Jansen Kill)

  Narrow ridges, plunging folds (vicinity of Great Vly)

* Post-glacial gorge (Austin Glen, northwest of Jefferson Heights)

* Reverse drainage (Roeliff Jansen Kill, etc.)

  Sand plain, glacial (east and south of Edgewood Club)

* Strike valley (occupied by Beaver Kill and Kaaterskill Creek)

  Topographic grain accentuated by glacial scour (generally north-south, throughout map)

* Water gap (High Falls)

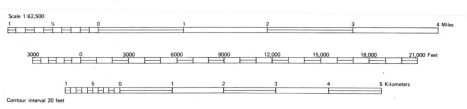

Contour interval 20 feet

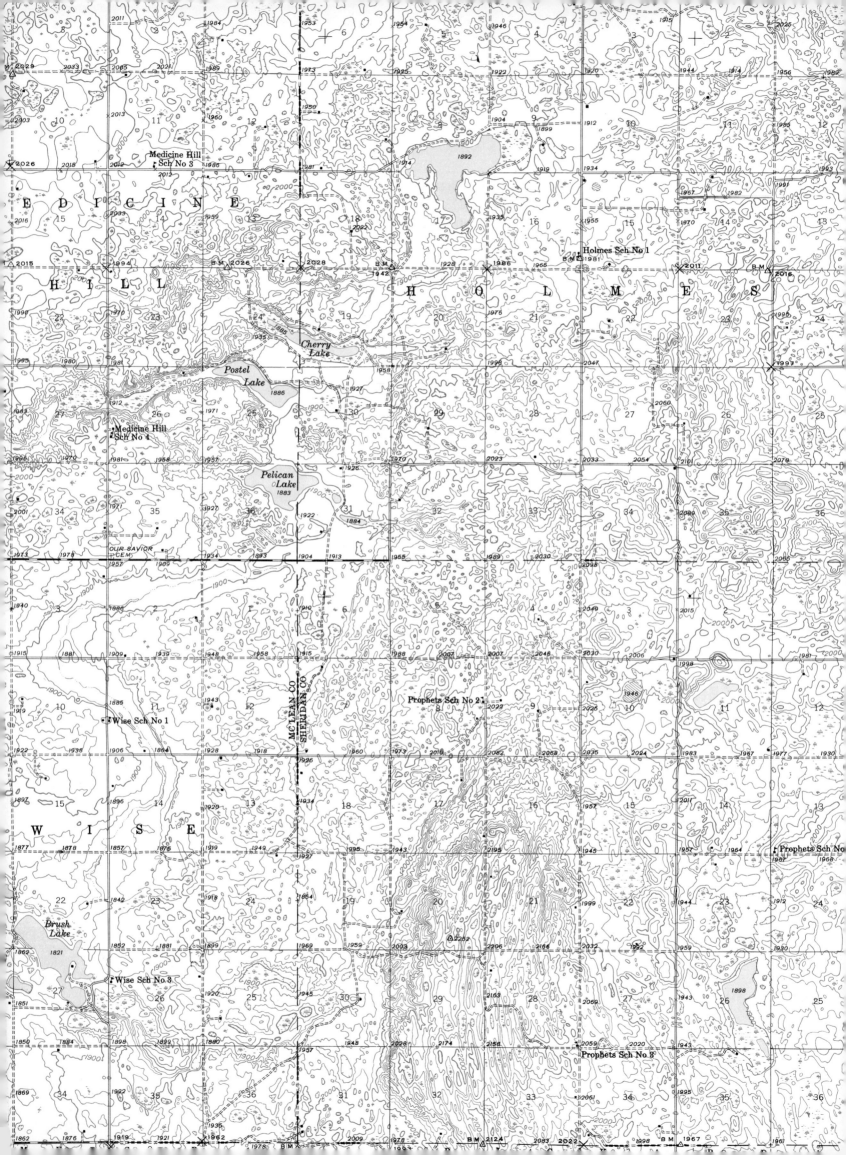

36 NORTH DAKOTA
PELICAN LAKE QUADRANGLE 1945

location

Central North Dakota
Great Plains Province
Missouri Plateau, Glaciated Section

physiographic features

* Coteau du Missouri region (entire map)
* End moraine topography (typical of Coteau du Missouri; covers major portion of map)
* Glacial outwash channel (through lakes area from elevation 1892 to Brush Lake)

　Kames (southeast corner of map)

　Kettles (over most of map)

　Lakes in kettle holes (along glacial outwash channel)

* Lobate washboard moraine, recessional (Sections 20, 21, 28, 29, 32, and 33 at south edge of map)

　Marsh or swamp (west of Postel Lake)

* Poorly integrated drainage (over entire map)

　Swell and swale topography (northwest portion of map)

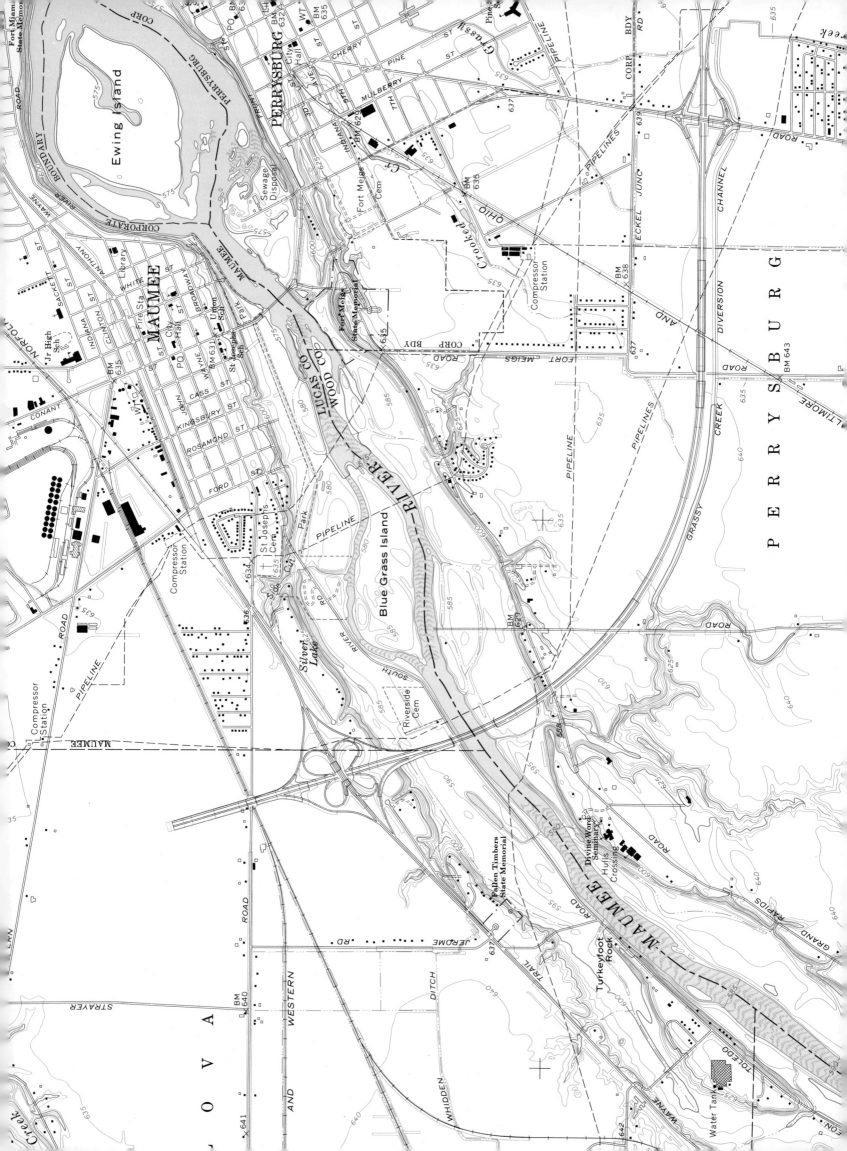

37

OHIO
MAUMEE QUADRANGLE 1951

location

Northwestern Ohio Near Lake Erie
Central Lowlands Province
Eastern Lake Section

physiographic features

* Abandoned channel (vicinity of Turkeyfoot Rock)

  Dissected lacustrine plain (over entire map)

* Glacial Lake Maumee (over entire area during period of continental glaciation)

* Meander scars or chutes (vicinity of Blue Grass Island)

* Plains remnant (isolated between Maumee River and abandoned channel at BM 339)

* Rapids (in Maumee River)

  River bluffs (along Maumee River)

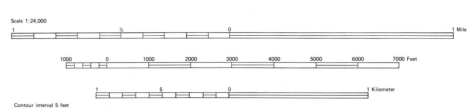

Scale 1:24,000

Contour interval 5 feet

38 OREGON
CRATER LAKE NATIONAL PARK MAP 1953

location

Southwestern Oregon
Cascade–Sierra Mountains Province
Middle Cascade Mountains Section

physiographic features

* * Ancient Mount Mazama (estimated height of this collapsed volcano, which was at one time the top of Crater Lake, is more than 14,000 feet)
* Benches formed by Nuees Ardentes deposits (along Sand Creek, Sun Meadow, and Sun Creek below Sun Meadow, etc.)
* Bluffs (Anderson Bluff, Scott Bluff, etc.)
* * Caldera of Ancient Mount Mazama (occupied by Crater Lake)
* * Cliffs (nearly the entire rim of caldera, notably Redcloud Cliff)
* * Collapsed volcanic cone (now occupied by Crater Lake)
* * Crater of Ancient Mount Mazama (now occupied by Crater Lake)
* Glacial notch on rim of caldera (Kerr Notch, etc.)
* * Lake with underground outlet (Crater Lake)
* * Nuees Ardentes deposits (*see* Bench above)
* * Parasitic cone (Red Cone, etc.)
* * Pumice Desert (high level plain, north part of map, formed by pumice)
* Radial drainage (from Ancient Mount Mazama)
* * Rim of caldera (crest surrounding Crater Lake)
* * Shield volcano capped by pyroclastic cone (Timber Crater)
* Spring (Anderson Spring, Lightning Spring, etc.)
* * Volcanic cone within caldera (Wizard Island)

Scale 1:62,500

Contour interval 50 feet

39

PENNSYLVANIA
ALTOONA QUADRANGLE   1920

location

West-central Pennsylvania
Appalachian Plateaus Province
Allegheny Mountain Section (also a small portion of Valley and Ridge Province)

physiographic features

* Allegheny Front (escarpment running in northeast direction at right angle to Bells Gap Run)
* Anticlinal valley (Sinking Valley)
* Anticline (at Sinking Valley)
* Disappearing stream (into Sinking Valley)
- Dissected plateau (northwest of Allegheny Front)
* Hogback (Brush Mountain)
- Sink holes (solution pits, vicinity of Culp)
* Structurally controlled dissected terraces (both sides of Sinking Valley)
* Structurally controlled ridges and valleys (east portion of map)
* Synclinal mountain (Brush Mountain)
* Synclinal valley (Canoe Creek)
- Water gap (Bells Gap, Riggles Gap, etc.)
- Wind gap (east of Kettle Reservoir where Kettle Creek beheaded Sinking Run)

note

This quadrangle joins and pairs with the Tyrone Quadrangle.

Scale 1:62,500

Contour interval 20 feet

40 PENNSYLVANIA
TYRONE QUADRANGLE 1929

West-central Pennsylvania
Valley and Ridge Province
Middle Section

physiographic features

* Anticlinal valley (Sinking Valley)

* Anticline (at Sinking Valley)

* Canoe-shaped mountain (around Canoe Creek connecting Canoe Mountain with Brush Mountain; there is an extension of Brush Mountain to the east of Plummer Hollow)

* Hogback (Tussey Mountain, Brush Mountain at Plummer Hollow, etc.; this end of Brush Mountain is very near Tyrone)

Sink holes (in Sinking Valley and near Oakland School)

* Structurally controlled dissected terrace (on flanks of Canoe Mountain and Brush Mountain)

* Structurally controlled ridge and valley (southwest portion of map)

* Synclinal mountain (Brush Mountain, Canoe Mountain)

* Synclinal valley (Canoe Creek)

* Water gap (near Tyrone on Little Juniata River at Plummer Creek; at Water Street on Frankstown Branch of Juniata River, just below the Kettle Reservoir on Kettle Creek)

note

This quadrangle joins and pairs with the Altoona Quadrangle. Imagine the pair of quadrangles side by side with Altoona on the left. It will then be clear that the land is complete.

Scale 1:62,500

Contour interval 20 feet

41

RHODE ISLAND
KINGSTON QUADRANGLE 1957

location

Coast of Rhode Island
Seaboard Lowlands Section
New England Province

physiographic features

   Beaches (on Block Island Sound)

   Drumloidal hill projecting through outwash plain (south part of map)

* End moraine (from area named "The Hills" westward through Bull Head Pond)

* Kames and kettles (from area named "The Hills" westward through Bull Head Pond)

   Kame terrace (south of Perryville)

* Kettles in moraine (in end moraine area)

* Lagoon (Potter Pond, with connection to the sea via the channel eastward from the letter "P" of Pond)

* Pitted outwash plain (south part of map)

* Ponds in kettles (The Hills)

note

Kingston, a small college town for which the quadrangle was named, lies north of this portion of the quadrangle.

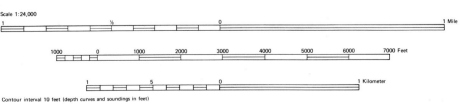

Contour interval 10 feet (depth curves and soundings in feet)

42 SOUTH CAROLINA
MULLINS QUADRANGLE 1947

location

Northeastern South Carolina
Coastal Plain Province
Sea Island Section
Carolina Bay Region

physiographic features

Bay (Fox Bay, Little Sister Bay, Big Sister Bay)

* Carolina Bay (origin variously attributed to solution of thin-bedded limestone, tidal swirls on falling coastline, or falling meteorites; origin of name, Carolina Bay, is not hydrographic, but very likely is botanical for Bay or Magnolia tree that is prevalent in area)

Coastal plain (terraced surface)

Flood plain (at Back Swamp)

* River bluff (bordering Back Swamp)

* Sand rims around Carolina Bays (Little Sister Bay, Big Sister Bay, etc.)

Scale 1:24,000

Contour interval 10 feet

43

SOUTH DAKOTA
SHEEP MOUNTAIN TABLE QUADRANGLE 1950

location

Near Southwestern South Dakota
Great Plains Province
Missouri Plateau, Unglaciated Section

physiographic features

* Badlands (over most of map)

  Badlands National Monument (north portion of quadrangle)

* Dissected plateau (well represented on slopes of Sheep Mountain Table and about Cedar Butte)

* Dunes and deflation hollows (southeast corner of area)

* Fine-textured topography (on slopes of Sheep Mountain Table, Cedar Butte, Pine Ridge, etc.)

  Intermittent drainage (entire map)

* Outlier (Cedar Butte)

  Structural terrace (southeast corner of area)

* Table (Sheep Mountain Table)

Scale 1:24,000

Contour interval 10 feet

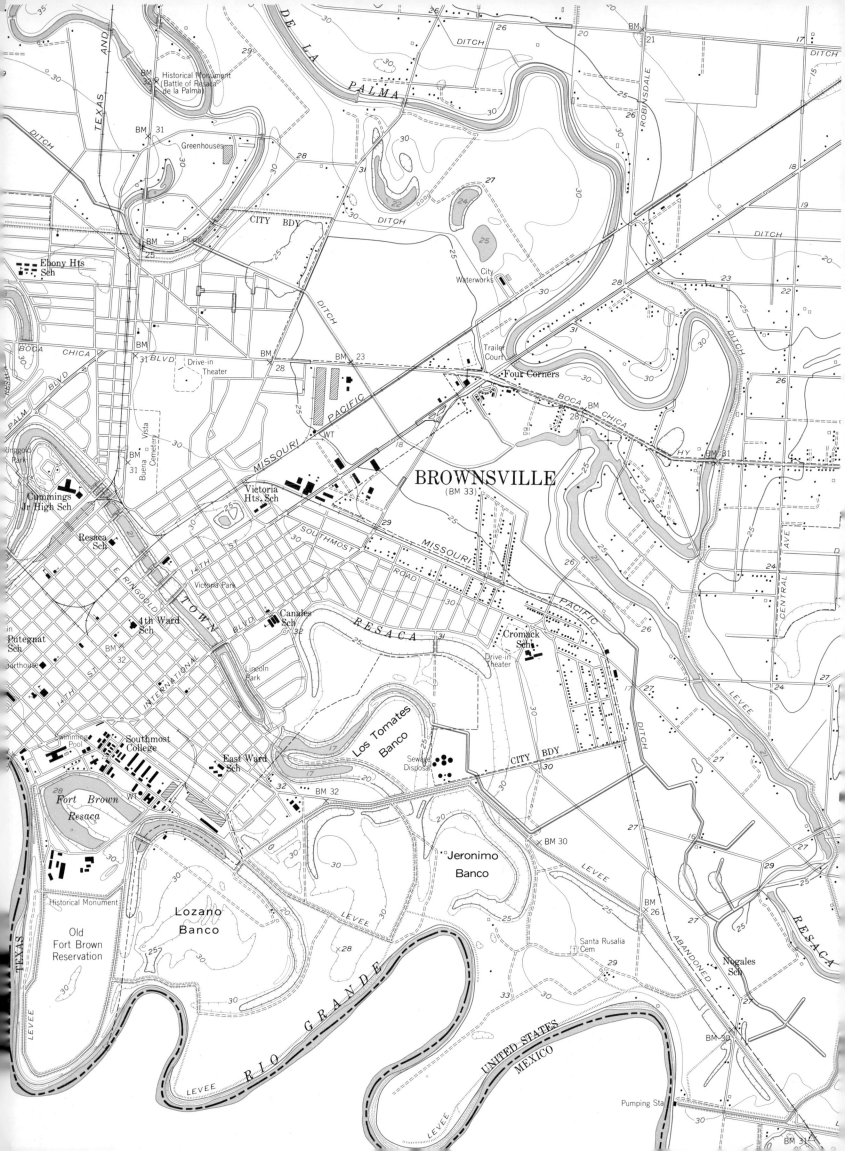

44

TEXAS
EAST BROWNSVILLE QUADRANGLE 1938-1955

location

Southern Texas at Lower End of Rio Grande Valley
Coastal Plain Province
West Gulf Coastal Plain

physiographic features

* Abandoned channels of Rio Grande (Town Resaca, also Resaca de la Palma where in 1846 was fought one of the first Mexican War Battles)

   Artificial levee (along north bank of Rio Grande, etc.)

   Cut-off meander (Los Tomates Banco, etc.)

* Delta of Rio Grande (entire map)

* Depression contours (numerous)

   International boundary (channel of Rio Grande)

   Irrigation canal (numerous and usually on artificial levees)

   Low relief (5-foot contour interval required)

* Meanders (Resaca de la Palma, Rio Grande, etc.)

* Oxbow lake (Fort Brown Resaca)

* Resaca (Resaca de la Palma, etc.)

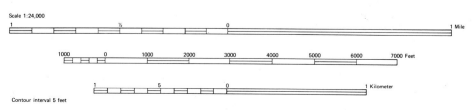

Scale 1:24,000

Contour interval 5 feet

45

TEXAS
GUADALUPE PEAK QUADRANGLE 1933

location

Near Southwestern Corner of Texas
Basin and Range Province
Block Mountain and Plain Sections

physiographic features

    Alluvial fans, coalescing (west side of map)

* Block mountains (Guadalupe Mountains)

* Bolson (Salt Basin)

* Cliff (northward from El Capitan to Bush Mountain)

* Depression contours (in northwest corner of map)

* Dissected upland (Guadalupe Mountains)

    Escarpments (northward from El Capitan to PX Flat)

    Fault scarp (southeast from Williams Ranch)

    Fault valley (east of Patterson Hills)

    Highest point in Texas (Guadalupe Peak—8751)

* Playa (Salt Lake)

* Salt Basin (historic for centuries; the source of salt for Indians)

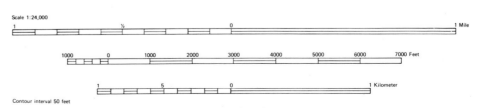

Contour interval 50 feet

46

UTAH
JORDAN NARROWS QUADRANGLE 1951

location

Northern Utah about 20 Miles South of Great Salt Lake
Basin and Range Province
Great Basin Section

physiographic features

* Abandoned meander of Jordan River (at Nash)

* Ancient cuspate bar with 5140-depression contour (at Point of the Mountain)

* Bonneville Shoreline (conspicuous on this map from Point of the Mountain along base of Sheep Mountain, generally between 5100 and 5200 contours)

  Dam (at Jordan Narrows; provides power)

  Irrigation canal (numerous)

* Lacustrine plain (above and below Jordan Narrows)

* Narrows (Jordan Narrows)

* Provo Shoreline (between 4700 and 4800, generally along the 4780 contour on this map, but less distinct than Bonneville Shoreline)

* Water gap (Jordan Narrows)

note

An ancient spit and ancient tombolo at the Bonneville Shoreline on the Jordan Narrows Quadrangle are west of and outside this area.

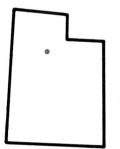

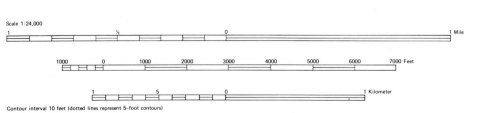

Scale 1:24,000

Contour interval 10 feet (dotted lines represent 5-foot contours)

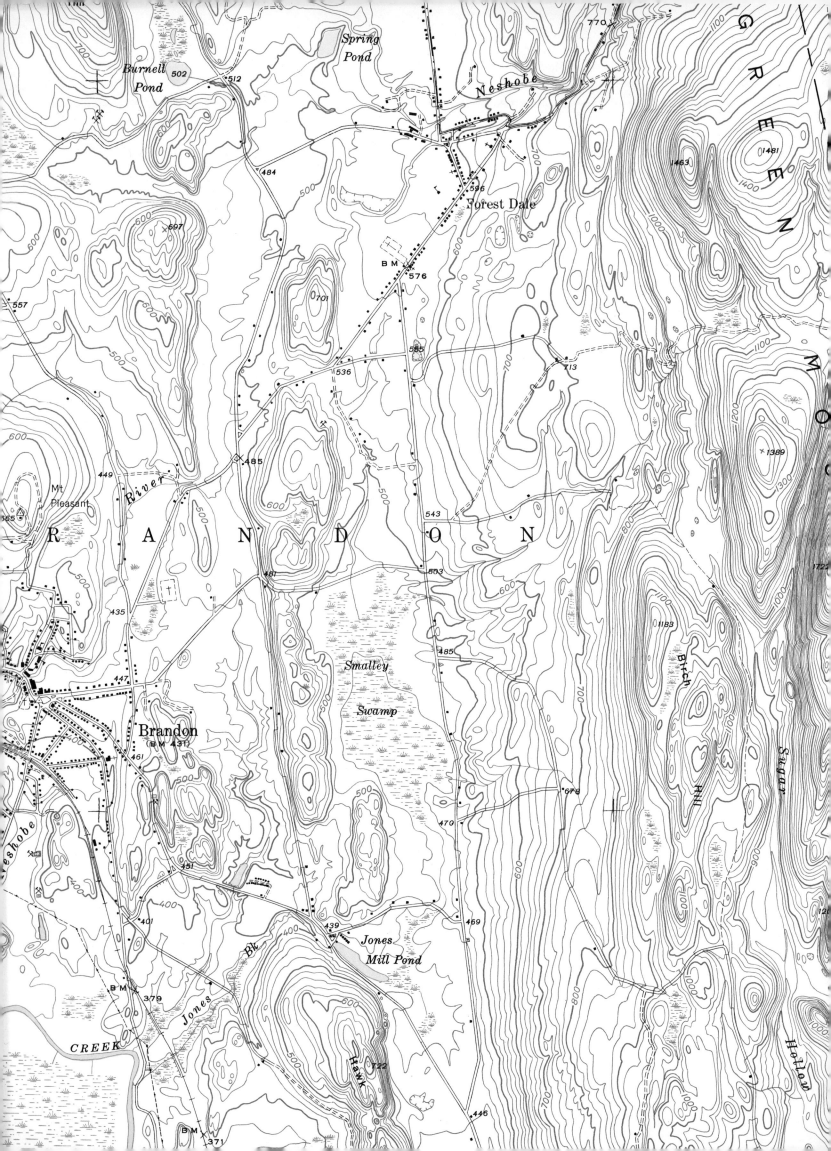

47

VERMONT
BRANDON QUADRANGLE 1946

location

Central Vermont
New England Province
Green Mountain Section

physiographic features

* Abraded bedrock hills (Hawk Hill, Birch Hill, etc.)

Hanging valley (around Jones Mill Pond)

Marble area (through upper center of map; underground, without surface signs)

* Obstructed drainage (Neshobe River adjacent to and north of Brandon)

Ridge and valley (most of map)

* Strike ridges, structurally controlled (Hawk Hill, northward; also Birch Hill)

* Strike valleys (on each side of Hawk Hill; also Sugar Hollow)

Swamp (Smalley Swamp, etc.)

Water gap (on Neshobe River near BM 485)

* West front of Green Mountains (rising east from Sugar Hollow)

Wind gap (north of Birch Hill)

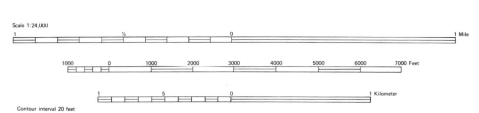

Scale 1:24,000

Contour interval 20 feet

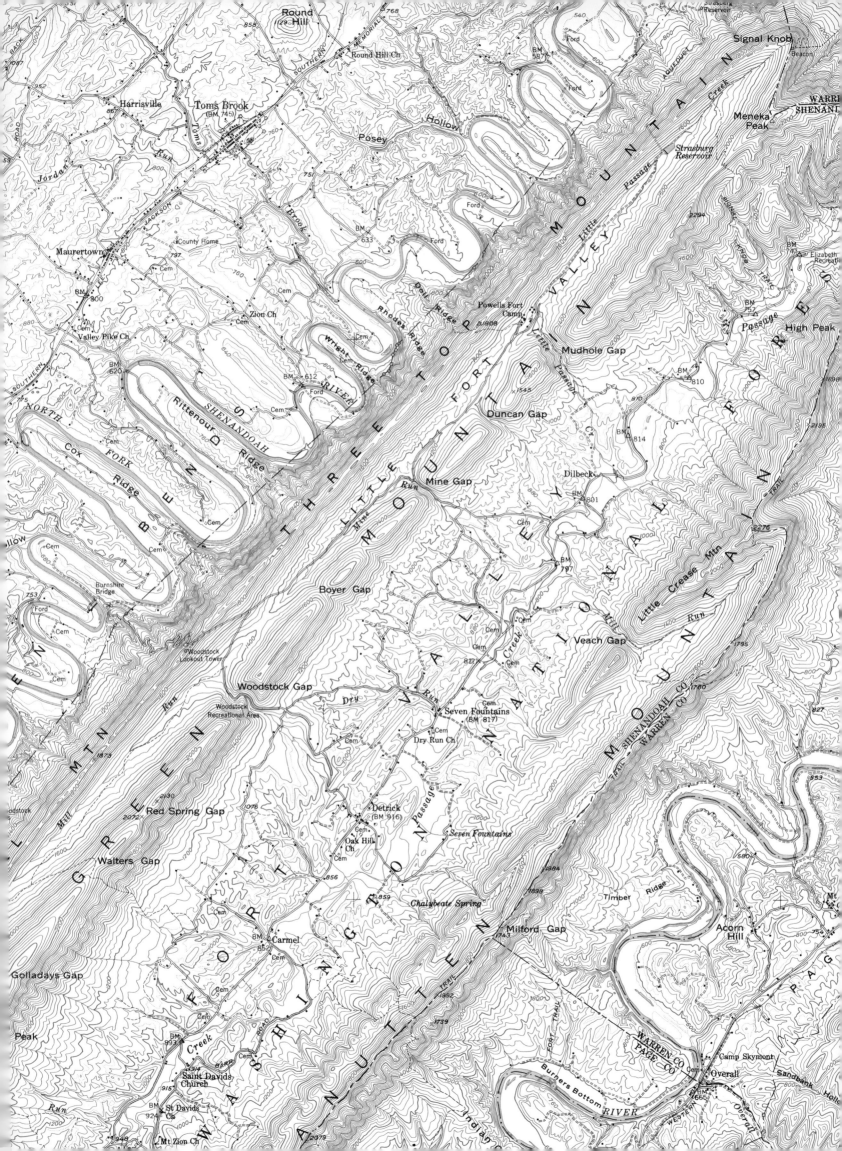

48 VIRGINIA
STRASBURG QUADRANGLE 1950

location

Northern Virginia
Valley and Ridge Province
Middle Section (also portion of Blue Ridge Province)

physiographic features

Accordant summits with sharp crests (center of map)
* Canoe-shaped mountain (around Little Fork Valley)
Dip slope (west slopes of Three Top Mountain, etc.)
* Entrenched meanders (North Fork Shenandoah River)
Escarpments (east slopes of Three Top Mountain, etc.)
* Meanders (North Fork Shenandoah River, South Fork Shenandoah River)
Rapids (South Fork Shenandoah River)
* S-shaped ridge (junction of Little Crease Mountain and Massanutten Mountain)
Strike valleys and ridges (eastern portion of map)
* Synclinal valley (Fort Valley, Little Fort Valley)
* Trellis drainage (Mile Run, Mine Run, etc.)
Water gap (Mine Gap, Veach Gap, etc.)
* Wide meander belt (North Fork Shenandoah River—an extraordinary example)
Wind gap (Boyer Gap, etc.)

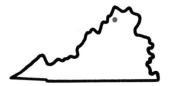

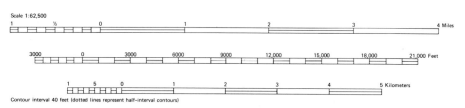

Scale 1:62,500
Contour interval 40 feet (dotted lines represent half-interval contours)

49 | WASHINGTON
MOUNT RAINIER QUADRANGLE 1924

location

Southwestern Washington (in Mount Rainier National Park)
Cascade-Sierra Mountains Province
Northern Cascade Mountains Section

physiographic features

* * Alpine topography (most of area)
* Cirque (Willis Wall, etc.)
* Cirque lake (Crescent Lake, Lake Ethel, etc.)
* * Col (Cayuse Pass, above Owyhigh Lakes, etc.)
* Falls (Marie Falls, Basaltic Falls, etc.)
* Glacial trough (White River)
* * Glaciers (more than on any mountain in the United States)
* * Hanging valley (Maple Creek above Maple Falls, Shaw Creek below Owyhigh Lakes, etc.)
* * High relief topography (more than 13,000 feet of this area between summit of Mount Rainier and mouth of Muddy Fork of Ohanapecosh River)
* Matterhorn (Pinnacle Peak)
* * Moraines, lateral and medial (at Cowlitz Glacier and Emmons Glacier)
* * Mountain peak, isolated (Mount Rainier)
* * Natural bridge (east of Independence Ridge)
* * Nunatak (St. Andrews Rock, Glacier Island)
* * Radial drainage on volcanic cone (on Mountain Rainier)
* Tarn (head of Lost Creek, etc.)
* * Volcanic cone and crater (Mount Rainier)
* Wind gap or pass (Chinook Pass, on Cascade Range)

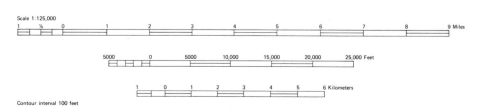

Scale 1:125,000

Contour interval 100 feet

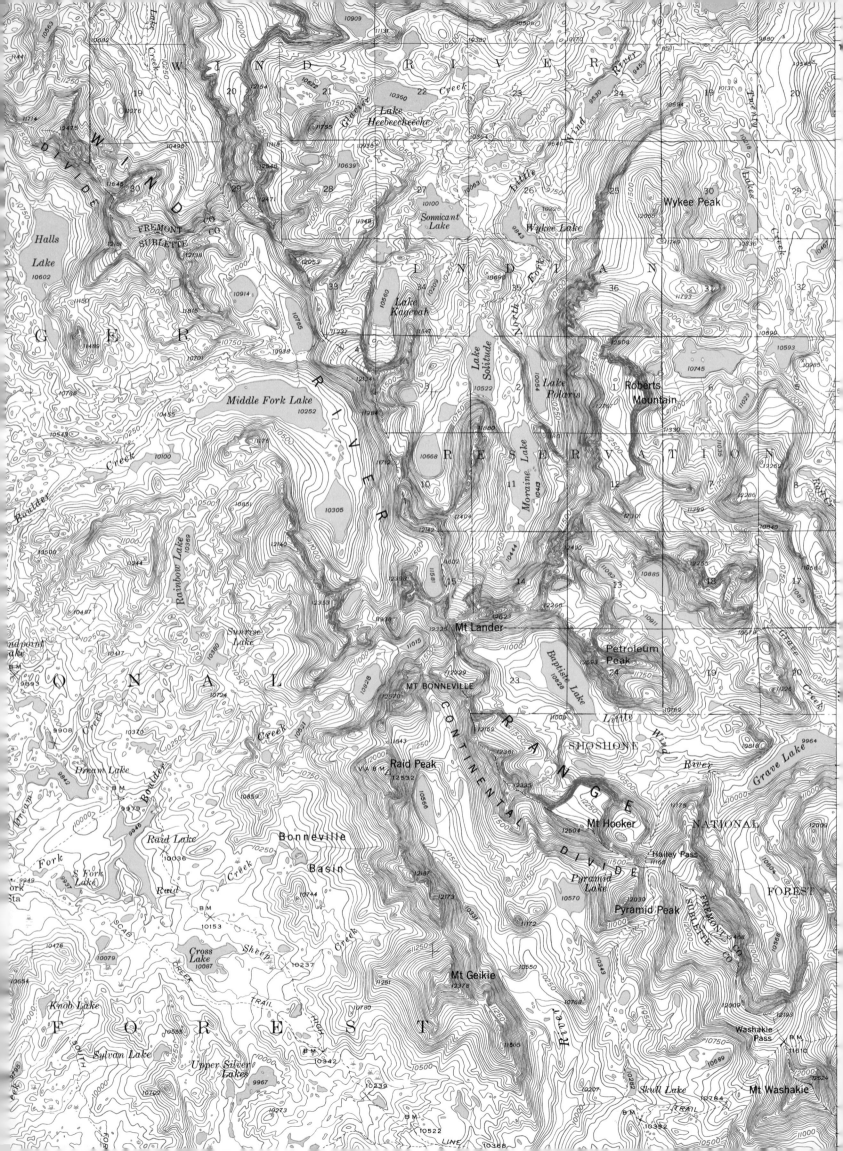

50

WYOMING
MOUNT BONNEVILLE QUADRANGLE 1938

location

West-central Wyoming
Middle Rocky Mountains Province

physiographic features

* Alpine glaciation (spectacular in east portion of map)

* Alpine topography (detailed beautifully on east portion of map)

* Arete (north and south of Mount Lander, etc.)

  Basin (Bonneville Basin)

* Cirque (above Lake Kagevah, north of Mount Lander, etc.)

* Col (Hailey Pass)

* Compound cirque leading to U-shaped valley containing moraine lake (head of North Fork of Little Wind River)

* Continental Divide (also the boundary between Fremont and Sublette Counties)

* Contrasting topography (east portion of map: spectacular Alpine glaciation; west portion of map: scoured by ice cap)

  Glacier (above Lake Kagevah, below Mount Lander, etc.)

  Hanging valley (Baptiste Lake at head of Little Wind River, etc.)

* Matterhorn (Mount Washakie, etc.)

  Mountain range (Wind River Range)

* Pater Noster Lakes (west portion of map, vicinity of Sunrise Lake, etc.)

  Structurally controlled linear drainage (west of Continental Divide)

* Subsummit erosion surface scoured by ice cap (west portion of map)

  Summit erosion surface remnants (scattered)

  Tarn (just east of Mount Hooker south of Baptiste Lake, etc.)

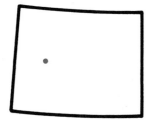

Scale 1:62,500

Contour interval 50 feet

115

GEOLOGIC MAPS COVERING TOPOGRAPHIC QUADRANGLES

LIST 1. U.S. Geological Survey geologic maps, published separately.*

Topographic quadrangle	Geologic Map	
Waldron, Arkansas	OM 192	Preliminary geologic map of the Waldron quadrangle and adjacent areas by J. A. Reinemund and Walter Danilchik, 1957. Scale 1:48,000
Juanita Arch, Colorado	GQ 81	Juanita Arch, Colo., geology by E. M. Shoemaker, 1955. Scale 1:24,000
Sonoma Range, Nevada	GQ 7	Mount Tobin, Nev., geology by S. W. Muller, H. G. Ferguson, and R. J. Roberts, 1951. Scale 1:125,000
	GQ 11	Winnemucca, Nev., geology by H. G. Ferguson, S. W. Muller, and R. J. Roberts, 1951. Scale 1:125,000
	MF 236	Preliminary geologic map of Humboldt County, Nev., by Ronald Willden, 1960. Scale 1:200,000
	GQ 15	Golconda, Nev., geology by H. G. Ferguson, R. J. Roberts, and S. W. Muller, 1952. Scale 1:125,000
Ship Rock, New Mexico	C 29	Preliminary geologic map of the Ship Rock and Hogback quadrangles, San Juan Co., N. Mex., by E. C. Beaumont, 1955. Scale 1:48,000
Mount Rainier, Washington	I 432	Geologic map and section, Mount Rainier National Park, Wash., by R. S. Fiske, C. A. Hopson, and A. C. Waters, 1964. Scale 1:62,500

LIST 2. U.S. Geological Survey book publications containing geologic maps.†

Topographic quadrangle	Geologic publication
Washington West, D. C.	Professional Paper 217. Configuration of the bedrock surface of the District of Columbia and vicinity, by N. H. Darton, 1950, plate 1. Scale 1:31,680
	Bulletin 967. District of Columbia, its rocks and their geologic history, by M. S. Carr, 1950, plate 6. Scale 1:31,680

*May be purchased from the U.S. Geological Survey, Washington, D. C. 20242. Prices of the public service publications are reasonable and well within the student's budget.

†Available publications may be purchased from the U.S. Geological Survey, Washington, D. C. 20242; other out-of-print publications are generally on file for reference at universities.

	Geologic Folio 70. District of Columbia, by N. H. Darton and A. Keith, 1901. Scale 1:62,500
	Water-Supply Paper 1776. Geology and groundwater resources of Washington, D. C. and vicinity, by Paul M. Johnston, 1964, plate 1. Scale 1:62,500
Warm Springs, Georgia	Water-Supply Paper 819. Warm Springs, by G. W. Crickmay and D. F. Hewett, 1937, plate 1. Scale 1:62,500
Menan Buttes, Idaho	Water-Supply Paper 818. Geology and water resources of the Mud Lake region including the Island Park area, by H. T. Stearns, 1939, plate 3. Scale 1:125,000
Thousand Springs, Idaho	Water-Supply Paper 774. Geology and ground-water resources of the Snake River Plain in southeastern Idaho, by H. T. Stearns, 1938, plate 5. Scale 1:62,500
	Water-Supply Paper 1475-P. Hydrogeology of stock-water development in southeastern Idaho, by R. F. Hadley, 1963, plate 29. Scale 1:500,000
Mammoth Cave, Kentucky	Bulletin 688. Oil fields of Allen County, by J. B. Hoeing, E. W. Shaw, and K. F. Mather, 1919, plate 10. Scale 1:250,000
Lynn, Massachusetts	Bulletin 839. Geology of the Boston area, by L. La Forge, 1932. Scales 1:62,500 and 1:1,250,000
	Bulletin 704. Geology of the igneous rocks of Essex County, by C. H. Clapp, 1921, plate 1. Scale 1:250,000
Crater Lake, Oregon	Professional Paper 3. Geology of Crater Lake National Park, by J. S. Diller and H. B. Patton, 1902, pp. 1-61, plate 6. Scale 1:142,560
Kingston, Rhode Island	Bulletin 1071-I. Surficial geology of the Kingston quadrangle, Rhode Island, by C. A. Kaye, 1960, plate 32. Scale 1:24,000
Guadalupe Peak, Texas	Professional Paper 215. Geology of the southern Guadalupe Mountains, by P. B. King, 1948, plate 3. Scale 1:48,000
Strasburg, Virginia	Water-Supply Paper 596-C. Ground water in the Ordovician rocks near Woodstock, by G. M. Hall, 1927, plate 7. Scale 1:48,000
	Professional Paper 484. Geomorphology of the Shenandoah Valley, Virginia and West Virginia, by J. T. Hack, 1965. Scale 1:250,000

LIST 3. Available state publications.

State departments	Address
Arizona Bureau of Mines	University of Arizona, Tucson, Ariz. 85721
California Division of Mines and Geology	Ferry Building, San Francisco, Calif. 94111
Kentucky Geological Survey	University of Kentucky, 307 Mineral Industries Building, Lexington, Ky. 40506
Maine Geological Survey	State Office Building, Room 211, Augusta, Me. 04330
Maryland Department of Geology, Mines, and Water Resources	214 Latrobe Hall, Johns Hopkins University, Baltimore, Md. 21218
Mississippi Geological Survey	Drawer 4915, Jackson, Miss. 39216
New York State Museum Bulletins	N. Y. State Education Building, Albany, N. Y. 12224
Rhode Island Development Council	State House, Providence, R. I. 03029

Topographic quadrangle	Geologic publication
Antelope Peak, Arizona	Arizona Bureau of Mines. Geologic map of Pinal County, by E. D. Wilson, R. T. Moore, and coworkers, 1959. Scale 1:375,000
Furnace Creek, California	California Division of Mines and Geology. Geologic map of California, Death Valley sheet, by C. W. Jennings, 1958. Scale 1:250,000
Point Reyes, California	California Journal of Mines and Geology, vol. 51, no. 3, Mines, resources, and mineral industries of Marin County, by W. E. Ver Planck, 1955, plate 4. Scale 1:125,000
Washington West, D. C.	Maryland Department of Geology, Mines, and Water Resources. Geologic map of Montgomery County and the District of Columbia, by E. Cloos and C. W. Cooke, 1953. Scale 1:62,500
	Maryland Department of Geology, Mines, and Water Resources. Prince Georges Co. and the District of Columbia, by B. L. Miller, 1911. Revised edition by E. Cloos and C. W. Cooke, 1957. Scale 1:62,500
Mammoth Cave, Kentucky	Kentucky Geological Survey, series no. 6. Geologic map of Edmonson Co., by J. M. Weller, 1929. Scale 1:62,500
	Kentucky Geological Survey, series no. 6. Geologic map of Hart County, by F. S. Withers, A. H. Sutton, G. R. Wesley, and D. H. Crabb, 1931. Scale 1:62,500
Katahdin, Maine	Maine Geological Survey, State Park Geological series no. 2. The geology of Baxter State Park and Mount Katahdin, by D. W. Caldwell, 1960, plate 1-C. Scale 1:140,000
Cumberland, Maryland	Maryland Department of Geology, Mines and Water Resources. Geologic map of Allegheny County, by H. L. Berryhill, Jr., G. W. Colton, G. W. de Witt, Jr., and J. E. Johnston, 1956. Scale 1:62,500

Philipp, Mississippi	Mississippi Geological Survey Bulletin No. 50. Tallahatchie County mineral resources, by R. R. Priddy and T. E. McCutcheon, 1942, plate 1 Scale 1:125,000
Catskill, New York	New York State Museum Bulletins 331 and 336. Geology of the Catskill and Kaaterskill quadrangles (in 2 parts), by R. Ruedemann and G. H. Chadwick, 1942, 1946. Scale 1:62,500
Kingston, Rhode Island	Rhode Island Development Council Geological Bulletin 9. Groundwater resources of the Kingston quadrangle, Rhode Island, by W. H. Bierschenk, 1956, plate 1. Scale 1:31,680

LIST 4. Publications of professional societies.

Name of society	Address
Geological Society of America	P.O. Box 1719, Boulder, Colo. 80302
Indiana Academy of Sciences	Indianapolis, Indiana
Economic Geology	121 Natural Resources Building, Urbana, Illinois

Topographic quadrangle	Geologic publication
Mobile, Alabama	G.S.A. Bulletin, vol. 61. Pleistocene history of coastal Alabama, by C. W. Carlston, 1950, pp. 1119-1130. Scale 1:250,000
Point Reyes, California	G.S.A. Memorandum no. 35. Geology of the coastal ranges north of the San Francisco Bay region, by C. E. Weaver, 1949, plate 6-13. Scale 1:62,500
Oolitic, Indiana	Proceedings of the Indiana Academy of Science, vol. 55. Buddha outlier of Mansfield sandstone, Lawrence County, by C. A. Malott, 1945, pp. 96-101, Fig. 2. Scale 1:375,000
Virginia, Minnesota	Economic Geology, vol. 25. Hydrothermal oxidation and leaching experiments, their bearing on the origin of Lake Superior hematite-limonite ores, by J. W. Gruner, 1930, pp. 837-868, Fig. 9. Scale 1:125,000
Mount Bonneville, Wyoming	G.S.A. Bulletin, vol. 66. Pleistocene geology of the southeastern Wind River Mountains, by G. W. Holmes and J. H. Moss, 1955, no. 6, plate 1. Scale 1:250,000

LIST 5. Other publications.

Topographic quadrangle	Geologic publication
Bright Angel, Arizona	Geologic Map of the Bright Angel quadrangle, Grand Canyon National Park, Ariz., by J. H. Maxson, 1961. Scale 1:48,000 Available from: Grand Canyon Natural History Association, Box 219, Grand Canyon, Ariz.
Mount Desert, Maine	The Geology of Mount Desert Island, Maine, with explanation and descriptive field guide, by Carleton A. Chapman, 1962, Map, Fig. 4. Scale about 1:130,000 Available from the author, Department of Geology, University of Illinois, Urbana, Ill.
Provincetown, Massachusetts	Harvard University, Museum Comp. Zoology Mem. 52. Geography and geology of the region including Cape Cod, the Elizabeth Islands, Nantucket, Marthas Vineyard, No Mans Land, and Block Island, by J. B. Woodworth, E. Wigglesworth, and coworkers, 1934. Scale 1:62,500
Virginia, Minnesota	16th International Geological Conference of the United States, 1933, Guidebook No. 27. Lake Superior region, by W. O. Hotchkiss and coworkers, 1933, Excursion C-4, plate 8, map by J. W. Gruner. Scale 1:125,000
Ennis, Montana	11th Annual Field Conference of the Billings Geological Society, 1960, West Yellowstone Earthquake area. Scale 1:250,000
Crater Lake, Oregon	University of California Department of Geologic Science Bulletin 25. Calderas and their origin, by H. Williams, 1941. Scale 1:187,500
Jordan Narrows, Utah	Brigham Young University Research Studies, Geologic Series, vol. 4, no. 4. Geology of the Jordan Narrows quadrangle, by G. G. Pitcher, 1957, plate 1. Scale 1:24,000

BIBLIOGRAPHY

Index to a Set of One Hundred Topographic Maps Illustrating Specified Physiographic Features, Assembled by William B. Upton, Jr., United States Department of the Interior, Geological Survey, Washington, D.C., 1955. (Portions of 50 of these maps are used in this book.)

This set of 100 topographic maps illustrates a wide variety of well-portrayed physiographic features. The set generally follows the physical divisions map of the United States and illustrates most of its sections or subdivisions. The reverse side of the *Index* to these maps outlines the physical divisions.

In the *Index*, the location of each topographic map is shown on the physical divisions map; the physiographic features they portray are listed and summarized. In the first list the features are arranged alphabetically by states under the quadrangles on which they appear; in the second list the features are arranged categorically along with the names of quadrangles on which they are most clearly shown. Within the 100-map set is a smaller group of 25 maps for those who are interested in a more limited study.

The *Index* is free to the public.

Glossary of Geology and Related Sciences, American Geological Institute, Washington, D.C., 1957

Dictionary of Geological Terms, Published by Dolphin Reference Books, Prepared by the American Geological Institute, Washington, D.C., 1957

Physiography of the United States, by Charles B. Hunt, W. H. Freeman Co., San Francisco, Calif., 1967

The author describes the physical geography, geology, and related phenomena, considering some ways in which they have shaped our history.

Physical Geology, by Chester R. Longwell, Richard Foster Flint, and John E. Sanders, John Wiley and Sons, Inc., New York, 1969.

Physical Geography, 3rd ed., by Arthur N. Strahler, John Wiley and Sons, Inc., New York, 1969.

GLOSSARY OF TERMS AND DEFINITIONS[†]

Abandoned Channel—Channel generally caused by tidal action or by flooding.

Abraded—Rock formations worn away by friction, generally by wind-blown sand or glaciation; occurs in bedrock hills.

Accordant Summit—Even-crested summit or peak along a ridge.

Aggraded—Slope or plain built up by deposits of sedimentary material (sand, gravel or clay) such as on an aggraded desert plain.

Alluvial Fan—Cone or fan-shaped deposit of alluvium made by a stream where it runs out onto a level plain; occurs generally where streams issue from mountains upon the lowland.

Alluvial Plain—Plain resulting from the deposition of alluvium by water.

Alluvium—Sediment laid down by streams on alluvial fans, plains, etc.

Alpine Glaciation—Erosion and deposition by glacial ice in mountainous terrain as exemplified in the Swiss Alps.

Alpine Topography—Detailed contouring of rugged lands carved by Alpine glaciation into aretes, cirques, cols, matterhorns, etc.

Amphitheater—Alpine landform, characterized by cliffs surrounding a basin or cirque, resembling a circular outdoor theater.

Ancient Beach Ridge—Cuspate bar, lake, river bluff, shoreline, spit, terrace, or tombolo (each a prehistoric landform that is still visible, notwithstanding geologic changes).

Ancient Lake Agassiz—See Glacial Lake Agassiz.

Ancient Lake Bonneville—See Bonneville Shoreline.

Ancient Lake Maumee—See Glacial Lake Maumee.

Ancient Mount Mazama—Named in 1896 by members of a mountain-climbing club calling themselves "Mazamas" (Spanish for mountain goats).

Anticlinal Valley—Valley that follows the direction or axis of an anticline.

Anticline—Folds in rock strata inclined like the ridge of a house. *See* Syncline.

Apex of Ozark Plateau—Summit of the Ozark Mountains

Arête—Acute, rugged, Alpine crest such as a sharp divide between two cirques; somewhat resembles a fish back. [‡]

Arroyo—Steep-sided, flat-floored valley of an intermittent stream; typical of the semi-arid region of the Southwest.

Asymmetric—Unsymmetrical, such as a dome without proper proportion of parts or folds.

Badlands—An almost impassable region, nearly devoid of vegetation, where erosion has cut the land into an intricate maze of ravines and pinnacles.

Bajada—Series of overlapping or coalescing alluvial fans.

Banco—An oxbow lake or meander cut-off from a river by an alteration of its course.

Bar—Offshore ridge of sand or gravel, at the mouth of a river or bay, submerged at high tide and usually parallel to the beach.

Baranca—Arroyo but shorter and narrower.

Barrier Beach—Long bar rising above high tide.

Barrier Beach, Raised—Pamlico and Pleistocene Beaches are examples.

Base Line—In Public Land Surveys an east-west surveyed line along an astronomic parallel that passes through the initial point. Any base line is established with special care and precision.

Basin—Valley, cirque, or amphitheater with a small surface outlet or with no outlet; also a bolson.

Basin and Range—Region dominated by faultblock mountains separated by sediment-filled basins.

Battered Sea Cliff—Seaside cliff formed and battered by wave action.

[†] Terms and definitions were adapted from, by permission, the *Dictionary of Geological Terms*, Copyright © 1957, 1960, 1962, American Geological Institute, Washington, D.C., All Rights Reserved, and similarly from the *Glossary of Geology and Related Sciences*, 2nd ed., Copyright © 1957, 1960, American Geological Institute, Washington, D.C., All Rights Reserved.

[‡] The ^ mark over an "e" in French is called "circonflexe."

Bay—Wide inlet along the shore of a sea or lake. *See* Carolina Bay.
Bayhead Bar—Bar built a short distance from shore at the head of a bay.
Baymouth Bar—Bar extending partially or entirely across the mouth of a bay.
Bayou—Small sluggish stream; abandoned river meander.
Beach—Gently sloping shore of a body of water.
Beach Ridge—Continuous mound of beach material behind the beach built up by wave or other action; may be ancient.
Bed—Floor or bottom on which any body of water rests.
Bedding—Beds or planes dividing sedimentary rocks, etc.
Bedrock Hill or Knob—Generally rounded solid rock at the surface, either exposed or overlain by unconsolidated material.
Bench—Flat or gentle sloping terrace with one or more steeply descending sides and one ascending side; may or may not be controlled by bedrock.
Bench Mark or BM—Relatively permanent object of natural or artificial material bearing a marked point whose elevation is above or below an accepted datum, generally mean sea level. The letters BM beside a location cross (X) with elevation to the nearest foot are found on Geological Survey maps.
Biscuit-Board Topography—Mountain ridge from the top of which a series of cirques has been scalloped or bitten by glacial action.
Blind Valley—Valley form in karst areas where a stream disappears into the closed end of the valley or where there is no stream at all.
Block Mountain—Mountain carved by erosion from uplifted earth blocks and bounded on one or both sides by fault scarps.
Blowout Dune—Term applied to the dune built from sand derived from a blowout trough.
Blue Ridge Front—Outer slope of the Blue Ridge Mountains. *See* Front.
Bluff—Any high bank or headland with a precipitous front.
Bolson—Large depression or valley having no surface outlet; a far-western feature.
Bonneville Shoreline—Still visible shoreline of enormous, pre-historic Lake Bonneville, Utah; was named for Benjamin DeBonneville, explorer.
Braided Stream—Stream flowing in several dividing and reuniting channels because of obstructions of sand or gravel deposited in the stream.
Breached Crater or Breached Volcanic Cone—Volcanic crater or cone where lava has burst through a wall and flowed out through the breach.
Butte—Conspicuous, isolated, small mountain or mesa with very steep sides.

Caldera—Large, basin-shaped, volcanic depression that is circular in form.
Canal—Artificial watercourse cut or built through land for navigation, mining, irrigation, etc.
Canoe-Shaped Mountain—Mountain of two narrow ridges, joined at the upper end, which surrounds a synclinal valley and resembles the end of a canoe.
Canyon—Steep-walled gorge. The term is generally used in the Western mountain and plateau regions; Eastern usage would be chasm. Canyon is occasionally spelled cañon with a "tilde" over the "n."
Cape—Point of land extending into a sea or lake; a headland.
Carolina Bay—Round or oval depression, generally marshy, with bay trees (which may be the key to the name); occurs along the Atlantic Coast. The origin is unknown but is variously attributed to solution of thin-bedded limestone, tidal swirls on a falling coastline, or falling meteorites.
Cave—Underground cavity or chamber that is large enough to be entered by man; generally produced by solution of limestone.
Centrifugal Drainage—Drainage pattern which originates and flows outward from the central, highest part of an area.
Channel—Deepest portion of a stream through which flows the main volume of water; a large strait.
Check Dam—Dam to divert a stream.
Cinder Cone—Conical hill or mountain formed by volcanic ash or clinker-like material around a volcanic vent.
Cirque—Deep steep-walled recess in a mountain caused by glacial erosion; a bowl-like Alpine feature.
Cirque Headwall—Steep wall-like cliff at the back of a cirque.
Cirque Lake—Small body of water occupying a cirque depression.
Clay Dune—Dune formed of clay instead of the usual sand dune.
Cliff—High steep face of rock; a precipice.
Coalescing Alluvial Fan—Union of alluvial fans made by neighboring fans merging laterally; a bajada.
Coastal Bar—*See* Bar.

Coastal Plain—Plain with its margin on the shore of a sea.
Coastal Terraces—Occur along the Atlantic Coast from Delaware to Florida (e.g., Okefenokee, Pamlico, Silver Bluff, and Wicomico); all are ancient and may contain mineral-bearing sands, generally on the seaward side.
Col—Saddle or gap between two peaks that is wider than an arete (not necessarily separating two cirques).
Colorado Pediment—Alluvial plain formed at the foot of the Front Range of the Rocky Mountains.
Compound Cirque—Cirque that merges with an adjacent cirque.
Compound Recurved Spit—Spit with its end strongly curved inward.
Concave Slope—Inclined slope of a hill that curves near its foot like the inside of a sphere. *See* Convex Slope.
Cone—Steep-sided pile of sand or rock with a small summit and circular base.
Continental Divide—Uplands or mountain chain that separates continental drainage basins.
Continental Glaciation—Results of the ice sheet which formerly covered the North-central and Eastern parts of the Continent; it transformed surfaces of regions and developed numerous physiographic forms.
Contour—Line of equal elevation shown in brown on a topographic map and based on mean sea level datum. (See the discussion in the Introduction; also see the "Topographic Map Symbols" chart for the intermediate contour symbol.)
Contour Interval—Difference in elevation value between adjacent contour lines. (See the discussion in the Introduction.)
Convex Slope—Slope curving down like the outside of a sphere; opposite of concave slope.
Corrasion—Mechanical erosion performed by moving agents such as glacial ice, wind, and running water.
Coteau Du Missouri—A strip of morainic plateau in the Dakotas, bounded on the west by the Missouri River and on the east by Central Lowlands.
Coulee—Steep-sided gulch which at times is of considerable length.
Cove—Popularly applied term to small areas of valleys or plains that extend into mountains or plateaus; also a small bay or open harbor.
Crater—Bowl-shaped volcanic depression with steep sides and of considerable size.
Crest—Eminence crowning a hill or mountain from which the surface dips downward in opposite directions.
Cuesta—Unsymmetrical ridge with one side gently sloping and the other a steep slope or escarpment.
Cumberland Front—Front of the Cumberland Plateau. *See* Front.
Cuspate Bar—Crescent-shaped bar uniting with the shore at each end.
Cut-Bank—Concave or undercut slope of a bank which is steeply eroded by a meandering stream.
Cut-Off—Formation of an oxbow by a meandering stream cutting through the neck of a horseshoe bend.
Cut-Off Meander—Oxbow formed by a cut-off.
Cyclopean Stairs—Massive stairs or steps of bare steep rock in a series between cirques and ice-gouged basins.

Datum, Elevation—Level reference elevation, usually mean sea level, used in topographic mapping and hydrography.
Datum, Horizontal—Basis on which geographic map coordinates are calculated. The North American datum is used by the Geological Survey.
Datum, Vertical—Base from which elevations (altitudes) are calculated. Mean sea level datum, used universally for mapping, is obtained by averaging hourly heights of the sea on the open coast or in adjacent waters having access to the sea, the average being taken over a considerable number of years.
Delta—Alluvial deposit at the mouth or end of a river.
Deltaic Channel—River channel flowing through a delta.
Dendritic Drainage—Pattern of drainage with tributaries joining a main stream somewhat as branches join the trunk of a tree.
Denudation—Process of washing away the covering of strata. *See* Strata.
Depression—Low place or hollow having no outlet for surface drainage.
Depression Contour—Contour with tick marks indicating a depression in the land. *See* Depression.
Deranged Drainage—Normal drainage patterns upset by continental glaciation.
Desert Plain—Large area of generally uniform slope in a dry barren region that is comparatively level.
Desert Topography—Contour lines representing land forms in dry desert regions which are less detailed than those in regions of normal rainfall due to lack of the strong erosional agent: moving water.
Diabase—Rock of basaltic (volcanic) composition.
Differential Erosion—The more rapid erosion of one portion of the Earth's surface compared with another.
Dike—Body of volcanic rock that cuts across the structure of adjacent rocks or massive rocks.
Dikes, Radial—Dike radiating from a central area.
Dip—Angle at which a stratum is inclined from the horizontal and at a right angle to the strike. *See* Strike.

Dip Slope—Slope angle at which any planar feature, such as a mesa, plateau, or stratum, is inclined from the horizontal; slope of the land surface which conforms to the dip of the underlying rocks.

Disappearing Stream—Surface stream which disappears underground.

Dissected—Land forms cut by erosion (subject to erosion are alluvial fans, domes, pediments, plains; including lacustrine plains, and till plains; plateaus, including glaciated plateaus and those of strong relief; terraces, including those structurally controlled; uplands; volcanoes; etc.)

Distributary Channels—Outflowing branches of a river or stream characteristic of a delta; may be abandoned in some cases.

Divide—Line of separation between drainage systems; highest summit of a pass or gap.

Dome—Roughly symmetrical upfold with the beds dipping in all directions, more or less equally, from a point.

Drainage—Means for effecting the removal of water by downward flow of streams and other agents.

Drift—*See* Glacial Drift.

Driftless—Being without glacial deposits or drift.

Drowned River or Coastline—Change in or near a coastline caused by the sinking of a large area which may drown river channels and other outlets and also create offshore islands; common to both coasts and inland waters; a submerged shoreline.

Drumlin—Elongated gravel hill, steep on one side, consisting of glacial drift deposits.

Drumloidal Hill—Hill with drumlin characteristics.

Dune—Mound, ridge, or hill, generally of wind-blown sand but sometimes of clay.

Dune, Lakeshore—Dune established on the windward side of a lakeshore.

Dune Topography—Intricate contouring of sand lands involving complex ridge and hollow patterns.

Elongate Fold—Lengthy fold of stratified rock. *See* Strata.

Encroachment of Drainage—Advance of a stream beyond its own channel into that of another and older drainage. *See* Stream Piracy.

End Moraine—Moraine marking the terminal position of a glacier; a terminal moraine.

Entrenched Meander—Meander eroded more or less deeply below the surface of the valley in which it was formed. *See* Meander.

Entrenched Stream—Narrow meandering trench of a stream cut in a flat-bottomed trough lying well below the general upland.

Erosion—Group of processes whereby rock or other material is loosened, dissolved, and removed from any part of the Earth's surface; probably the most effective method of altering land forms; includes corrasion, denudation, solution, transportation, and weathering. Moving ice (glacial), running water, waves, and winds are the principal erosion agents.

Erosional Remnant—Hill or mountain outlying from the mesa, plateau, or upland from which it was separated by erosion.

Erosion Surface—Eroded or dissected surface.

Escarpment—Steep face at the abrupt termination of a highland or plateau; a scarp. The eastern escarpment of the Sierra Nevada is an immense example.

Esker—Sinuous ridge of glacially accumulated gravel having the general direction of drainage in a former ice sheet.

Eskeroid Topography—Topography in which the portrayal of lengthy meandering eskers are symbolized excellently by contours.

Esplanade—Bench or mesa on a deeply dissected plateau.

Estero—Estuary.

Estuary—Lower course of a stream or river in which the tide ebbs and flows.

Faceted River Bluff—River bluff which has been steeply beveled by erosion.

Faceted Spurs—Ends of spur ridges which in a glaciated valley have been steeply beveled by glaciation, stream erosion, or faulting.

Facet—Irregularly scalloped rock surface.

Fall Line—Line characterized by waterfalls in a river; line drawn through falls showing the contact between slope and flatland.

Falls—Cascades, cataracts, or waterfalls.

Fault—Fracture or displacement of a rock formation which may be an inch or many miles long.

Fault Block Mountain—Land mass bounded on two opposite sides by faults.

Fault Scarp—Cliff formed by a fault.

Fine-Textured Topography—Intricate detail well represented by contours in such regions as badlands, sand dunes, and in areas of karst topography.

Finger Lake—Long, narrow, glaciated rock basin occupied by a long lake.

Fiord—Long, deep arm of the sea, occupying a channel with high, steep walls.

Flat—Area of level land much smaller than a plain or strata.

Flatiron—Steep, triangular, sloping type of hogback ridge, often occurring in a series on the flanks of a mountain.

Flood Plain—Flat river valley, built of recent sediments, which is subject to inundation when the river overflows its banks.

Flowing Well—Well from which water or oil flows without pumping.

Fold—Bend in rock strata.

Folded Mountain—Mountain formed on a large fold.

Folds En Echelon—Series of parallel folded ridges or mountains of similar size and shape.

Front—Outer slope of a plateau or mountain range rising above a plain.

Gap—Notch or pass in a mountain ridge.

Glacial Drift—Sediment, deposited by a glacier, predominantly of glacial origin (generally continental).

Glacial Gorge or Spillway—Narrow passage draining continental glaciated areas.

Glacial Lake Agassiz—Ancient lake named in 1879 for a famous geologist, Louis Agassiz.

Glacial Lake Bed or Bottom—Bed or bottom owing its existence to glacial scouring action or to surrounding deposits of drift.

Glacial Lake Maumee—Origin of this name is an old Indian village known as Omaumeeg.

Glacial Linear Ridge—Ridge resembling an esker in content but being either gently curved or straight in lineation.

Glacial Lobe—Tongue-like projection from the main mass of a continental glacier.

Glacially Deepened Trough or Valley—See U-Shaped Valley.

Glacially Modified or Rounded Hill—Hill partially changed in form or shape by continental glaciation.

Glacially Scoured and Plucked Hill—Hill eroded by glacial action removing projecting rocks and other surface irregularities.

Glacially Scoured Bedrock Basin—Basin scoured by glacial action and generally occupied by a lake.

Glacial Outwash Channel—Channel formed in the drift outside of the active glacier ice.

Glacial Trough—See U-shaped Valley.

Glacier—Mass of ice, with definite limits, moving in a definite direction and originating from compaction of snow by pressure; may or may not contain varied quantities of gravel, etc.

Glacier Wall—Cirque headwall.

Glen—Narrow steep-sided passage.

Gorge—Deep narrow passage with precipitous rocky sides.

Gravel—Accumulation of water-worn pebbles (not to be confused with artificial gravel cut for building material).

Gullied Stream Channel—Broad bottom of a stream containing a deeply eroded narrow channel.

Gully—Small ravine.

Gut—Narrow channel or connecting passage of water.

Half-Interval Contours—Contours supplementing the designated contour interval.

Hanging Valley—Valley whose floor is higher than the valley or shore to which it leads or falls.

Harbor—Shelter for ships that is generally reached through an inlet; varies in size up to a bay.

Headland—High steep-faced promontory extending into the sea; may be a cape, point, or promontory.

Headland, Truncated—Headland terminating abruptly as if broken off.

Headwater Swamp—Marshy area providing water for a stream or lake.

Highland—Mountainous or plateau-like land mass.

High Level Plain—Plain in a mountainous area.

High Relief—Steep area with an abrupt vertical change from a low to a high altitude.

High Water Table—Zone where the upper surface of subsurface saturation is near or at ground level.

Hill—Natural elevated surface smaller than a mountain.

Hogback—Ridge produced by highly tilted strata, anticlinal, decreasing in height at both ends until it runs out.

Homoclinal Ridge—Ridge that slopes gently on one side and steeply on the other; cuesta.

Hook—End of a spit turned toward the shore.

Hot Spring or Hot Water Well—Thermal spring or well with water temperature above 98 degrees.

Ice Cap—Small ice sheet or glacier spreading out from a center.
Ice Carved—Ridge or other feature abraded by glacial action.
Ice Cave—Cave in which ice forms and persists through most of the year.
Inselberg—Prominent hill, usually one of a group, rising abruptly from a desert plain.
Integrated Drainage—Drainage that has become incorporated into a master drainage system.
Intermittent—Lake, stream, or drainage that contains or flows water less than half the year.
Intermont Basin—Basin lying between mountains.
Intermorainal Lowland—Low area between moraines.
Island—Land mass, not as large as a continent, surrounded by water.

Joint—Fracture in rock, generally vertical or transverse to bedding, along which no appreciable movement has occurred.

Kame—Conical hill or short irregular ridge of glacial drift deposited in contact with glacier ice.
Kame Terrace—Terrace of glacial sand and gravel deposits.
Kames and Kettles—Area of kames interspersed with kettles.
Karst Topography—Irregular topography abounding with large and small sinks, streamless valleys, etc., and occurring in limestone regions.
Kettle—Depression in glacial drift made by the melting of glacier ice buried in the drift; often contains a lake or pond.
Kettle Hole—Large bowl-shaped depression in glacial drift which was the resting place of a huge glacier ice mass before final melting.
Kill—A Dutch word for creek, stream, or channel used in New York State.
Klippe—Isolated rock mass separated from the underlying rocks by a fault; may be an erosional remnant or may have moved into place by gravity sliding.
Knob—Rounded hill or mountain, local in the South; also a hill of glacial origin in northern Middle West.
Knobs and Kettles—Hills and depressions in areas of glacial drift.

Lacustrine Plain—Minor type of plain occupying the basin-filled bed of an ancient extinct lake.
Lagoon—Body of shallow water with a restricted connection to the sea.
Lake—Any standing body of inland water that is larger than a pond and may be of considerable size.
Landform—One of the multitudinous features (hills, valleys, plains, etc.) which together make up the surface of the Earth. *See* Physiographic Feature.
Landslide—The sliding down a slope of a mass of rocks or earth which has become loosened or detached; may cause a dam and lake in the stream below.
Lateral Moraine—End moraine built along the lateral margin of a glacier and beyond the main mass of ice or drift.
Laurentian—Granite formation of an early geologic age.
Laurentian Divide—Summit of the Laurentians which are known as the Canadian Highlands or Laurentian Shield.
Lava—Molten rock issued from a volcano or an earth fissure; the same material solidified by cooling.
Lava Flow—Solidified mass of rock formed when a lava stream congeals.
Limestone—Bedded sedimentary deposit consisting chiefly of calcium carbonate.
Linear—Straight or gently curved physiographic feature such as a ridge formed either by glaciation or controlled by bedrock structure.
Lobate Washboard Moraine—Projection of a body of rough glacial drift beyond the main mass of drift.
Lobe—*See* Glacial Lobe.
Loess—Deposit of silt, very fine sand, or clay. The term originated in the Rhine Valley in about 1821.
Longitudinal Valley—Valley having a direction parallel to the strike of the adjacent ridges.

Map—Representation on a plane surface at an established scale of the physical features, natural or artificial, of a part or whole of the Earth by means of symbols with the method of orientation, such as North, indicated.
Map Scales—Scale expressing the size relationship between features shown on the map and the same features on the Earth's surface. This relationship is usually expressed as a ratio or fraction such as 1:62,500 or 1/24,000. The numerator 1 represents map distance; the denominator 24,000 represents ground distance. Thus the scale 1:24,000 states that any unit, such as 1 inch or 1 foot, on the map represents 24,000 of the same units on the ground; and it can be expressed as 1 inch represents 24,000 inches or 2000 feet on the ground. A scale of 1:24,000 is a large scale map; it shows greater detail but covers less area than a so-called

small scale map of 1:62,500. Thus it takes four 1:24,000 scale maps containing some 60 square miles to produce, after reduction, a 1:48,000 scale map covering about 240 square miles.

The larger scale quadrangles, such as 1:24,000, are of 7½-minute size. The small scale quadrangles, such as 1:62,500, 1:125,000, or 1:250,000, are of 15-minute, 30-minute, or 1-degree size, respectively.

Marine—Of, belonging to, or caused by the sea.
Marine Terrace—Plain of marine denudation, erosion, etc.
Marsh—Area of flat wet ground, usually covered with rank vegetation but without trees.
Marshy Divide—Wide divide or pass occupied by a marsh.
Matterhorn—Sharp, hornlike, or pyramid-shaped mountain peak resembling the Swiss peak of that name.
Meander—One of a series of regular and looplike bends in a meandering stream.
Meander Belt—Zone between two lines that is tangent to all outer meander loops.
Meander Core—The central hill encircled by the meander.
Meander Patterns—Patterns formed by abandoned meanders.
Meander Scar—Crescentic cut on the valley land bordering a stream: formerly an outer loop of a meander.
Medial Moraine—When two mountain glaciers unite, their coalescing lateral moraines form a medial moraine.
Meridian—Line of longitude; true north line.
Meridian (principal)—In public land surveys, a line extending north and south, passing through the initial point, along which township lines, section lines, etc., are measured.
Mesa—Tableland or flat-topped mountain bounded on at least one side by an escarpment.
Migrating Divide—Divide that is moving as the alignment of the physiographic feature which it follows is changed by erosion.
Mineral Spring—Spring whose water contains large quantities of mineral salts.
Monadnock—Residual hill or mountain standing above a nearly flat undulating plain or peneplain. This term was derived from the well-known Monadnock Mountain in New Hampshire.
Monolith—Prominent dome or knob composed of a single block of stone.
Morainal Lake—Lake owing its existence to the blockade of a valley or drainage course by glacial drift.
Moraine—Drift deposited chiefly by direct glacial action and independent of control by the surface on which it lies.
Morainic Plateau—Plateau covered by glacial drift.
Morainic Topography—Very detailed and irregular land surface, composed of drift, pertaining to, forming, or formed by a moraine.
Mountain—Land mass that is considerably elevated above the adjacent country (usually more than 2000 feet in altitude) and is larger than a hill.
Mountain Peak—Mountain whose summit reaches a sharp point.
Mountain Range—Chain of mountains.

Narrows—Narrow passage, generally for water.
Natural Bridge—Natural stone arch spanning a valley or ravine.
Natural Levee—Long alluvial ridge, built up on either side of a stream in floodtime by stream-carried silt, in a valley plain.
Neck—Narrow strip of land such as an isthmus or long hill; narrow band of ocean water.
Nonintegrated Drainage—Drainage without a master drainage pattern.
Notch—Deep pass or gap in a ridge or mountain; short canyon.
Nuees Ardentes Deposit—Highly heated mass of gas-charged lava ejected nearly horizontally from a volcano and flowing swiftly on its course down an outer slope.
Nunatak—Isolated hill or peak which projects through the surface of a glacier.

Obstructed Drainage—Drainage that is hindered or blocked because of either glaciation or landslide.
Oil Field—District containing a proved subterranean store of petroleum with economic value.
Okanogan Highlands—Area named for an Indian tribe.
Old—Maturity reached by a physiographic feature when erosion is at least three-quarters complete.
Outlier—Portion of any stratified area, such as a plateau, which lies detached, having been separated by denudation.
Outwash—Drift deposited by melting streams beyond active glacier ice; similarly, outwash-filled channels and outwash terraces.
Oxbow—Crescent-shaped lake or swamp area formed in a river bend which has been isolated by a change in the course of the river.

Pamlico Beach, Pamlico Shoreline, Pamlico Terrace—Ancient land forms. The name Pamlico was taken from the Pamlico River in North Carolina. The Pamlico Shoreline in Northern Florida is mineral bearing on the seaward side.

Parallel Drainage—Drainage where a number of adjacent streams flow parallel, or nearly so, to one another.

Parasitic Cone—Cinder cone formed on the flanks of a volcano.

Pass—Gap or saddle in a mountain ridge.

Pater Noster Lakes—Series of small connecting lakes in a glacial stairway that somewhat resembles a string of Pater Noster Beads.

Pediment—Gently inclined rock surface between the base of a mountain and a valley or basin bottom.

Peneplain—A land surface worn down by erosion to a nearly flat or undulating plain.

Physiographic Feature—Landform of which genesis and evolution are known.

Physiography—Study of the genesis and evolution of landforms.

Piedmont—Lying or formed at the base of a mountain; as a Piedmont glacier or as a Piedmont alluvial plain.

Piedmont Reentrant—Indentation of the Piedmont into the mountain range.

Pitted Outwash Plain—Plain composed of outwash containing pockets of ice, thus leaving holes in surfaces.

Placer Deposit—Mass of gravel or sand resulting from the crumbling and erosion of solid mineral-bearing rocks, generally containing some particles or nuggets of gold or of other valuable minerals.

Plain—Any extent of level or gently sloping land.

Plains Remnant—Portion of a plain left isolated by the erosive action of a stream or streams.

Plateau—Elevated level or nearly level area that ends abruptly in a steep slope on one or more sides.

Platform—Area of thinner sediments adjoining a geosynclinal wedge of thicker equivalent beds. In this particular case, geosynclinal means subsidence as exemplified by the Tonto Platform in Grand Canyon area.

Playa—Shallow central basin of a desert plain in which water gathers after a rain and then is evaporated.

Pleistocene—Geologic subdivision known as the Glacial epoch.

Plunging Fold—Steeply dipping strata.

Point—Projection from the shore of a body of water; portion of a front from which a wide area can be viewed.

Pond—Body of water that is smaller than a lake.

Ponds in Kettles—See Kettles.

Poorly Integrated Drainage—Region where undrained depressions have not become incorporated into a drainage system.

Prograded Shore—Seaward advance of the shoreline.

Projection—Geometric figure at any selected scale, plotted from precisely calculated coordinates, on which a map can be constructed having relatively precise distances and directions as compared to the curved Earth surface. The Geological Survey quadrangles generally have been mapped on the polyconic projection, nearly tangent to the Earth, on which distances are nearly the same as on the spheroidal Earth.

Promontory—Headland with a bold termination; a cape on the sea coast but also inland on the steep point of a plateau.

Public Land Subdivisions—Divisions originally surveyed and established by the General Land Office, now the Bureau of Land Management. These Public Surveys have divided most of the public domain into townships and sections with established corners of a permanent nature. There are 36 sections or square miles in a township.

Pumice—A light weight, porous form of lava.

Pyroclastic—General term applied to volcanic material explosively ejected from a volcanic vent.

Quadrangle—Name applied to the standard topographic map of the Geological Survey. Each quadrangle forming a part of the nation's topographic map coverage is designated by the name of a city, town, or prominent natural feature such as a mountain or lake. The maps are called quadrangles because they are not quite rectangles, being narrower across the north boundary than the south. Features along the edges of a quadrangle fit exactly with those along each adjacent quadrangle.

Quarry—Open working for the removal of building stone such as granite, limestone, and marble.

Radial Drainage—Drainage where streams radiate from a central area such as on a volcano or monadnock.

Raft—Jam of logs and trees falling from caving banks of a river.

Raised Ancient Beaches—Beaches uplifted in preceding geologic times.

Range—Chain of mountains.

Range Line—North-south boundary of a township.

Rapids—Part in a stream where the water moves much more swiftly than in adjacent parts.

Relief—Difference in elevation between high and low points of a land surface.

Remnant—See Erosional Remnant.

Resaca—Abandoned river channel or meander, usually containing water and of considerable length.
Resistant Rock Ridge—Ridge whose top stratum has qualities that resist erosion.
Reverse Drainage—Tributary drainage flowing toward a point that is further upstream than the logical junction with the main stream.
Ridge—Long, prominent land mass, generally with steep sides.
Rift—Very extensive fault such as the San Andreas Rift in California.
Rim—Rimrock forming the precipitous boundary of an elevated area.
River—Large stream where water flows from higher to lower levels, often flowing as far as the sea.
River Bluff—*See* Bluff.
River Mouth—Exit of a river into another river, lake, or sea.
Rock—Any consolidated, relatively hard, naturally formed mass of mineral matter.
Rock Awash—Partially covered rock projection in a stream or body of water.
Rock Terrace—Terrace found on the side of a valley cut in horizontal beds of unequal strength. The terrace is formed by the strong beds which are worn back less rapidly than the weak beds above and below them.

Salt Basin, Lake, or Spring—Basin, lake, or spring containing a predominating amount of sodium chloride.
San Andreas Rift—A famous fault 600 miles long extending in California from a point north of San Francisco southward through the state to the Mexican Border.
Sand Bar—*See* Bar.
Sand Dune or Hill—*See* Dune.
Sand Plain—Plain partially or completely covered by sand.
Sand Rim—Narrow zone of sand such as is found around the Carolina Bays.
Sand Scroll—Sand bars forming symmetrical patterns along a beach.
Sand Spit—Spit formed of sand. *See* Spit.
Scale—The ratio or proportion which a distance on a map bears to that same distance on the ground.
Scarp—Escarpment.
Scoured—Eroded by glaciation or by moving water.
Sea Cliff—Cliff formed by wave action.
Sea Level—Mean sea level is the vertical datum on which elevations and altitudes are based, particularly on the topographic maps of the United States Geological Survey.
Sea Stack—Small rock projecting above the sea near the shore.
Section—Portion of land 1 mile square; unit into which the public domain of the United States has been divided; the smallest of the physical divisions of the United States.
Section Corner—Corner at an extremity of a section boundary marked by a more or less permanent object. Corners are shown on a Geological Survey map by a vertical cross (+) in red or black.
Shield Volcano—Large, broad volcano with gently sloping sides.
Shorelines of Ancient Lakes—Bonneville and Provo for example.
Silver Bluff—Name of a limestone sea cliff near Miami.
Sink—Depression, generally of larger dimensions than a sink hole, evidenced by depression contours on a topographic map.
Sink Hole—Depression, generally in a limestone region, draining into a subterranean passage; often contains a lake or pond.
Sinuous—Winding divide; typical of an esker.
Slip-Off Slope—Sloping surface opposite the steep or undercut slope at a bend of a river or stream.
Slough—Place of deep mud; a mire that is generally filled by a sluggish stream.
Spit—Small point of land or sand projecting from the shore into a body of water. When recurved, the end of a spit is more or less strongly curved inward.
Spring—Place where water flows naturally from soil or rock.
Spur—Subordinate ridges extending from the crest of a mountain, like ribs from the vertebral column.
Square Mile—Area of land generally with four sides, each 1 mile long; any land area containing 640 acres.
Stack—Island or rock entirely removed from a headland by wave action and weathering.
Strand—Beach.
Strata—More than one sedimentary bed of rock; plural of stratum.
Stream—Any body of flowing water or other fluid.
Stream Piracy or Capture—Diversion of the upper part of a stream by the growth of another stream.
Strike—Course or bearing of the outcrop of an inclined bed or structure on a level surface that is perpendicular to the direction of the dip.
Stripped—Any relatively smooth surface from which sediment or rock debris has been removed.
Structural—Pertaining to the geologic structure such as a structural valley.

Superposed Stream—Stream with a present course which was established on young rocks burying an old surface. This course was maintained as the stream cut down through the young rocks into the old surface during uplift.
Swale—Short smooth ravine, ungullied.
Swamp—Low flat area that is partially or entirely covered with water and frequently with scattered trees.
Swell and Swale Topography—Topography of a ground moraine having low relief and gentle slopes.
Syncline—Fold in rock strata which dips inward from both sides toward the axis, as a trough; opposite in form of an anticline.

Table—Flat elevated area; a mesa, mountain, or plateau.
Tarn—Small lake or pool occupying an ice-gouged basin in a cirque.
Terminal Moraine—Moraine formed across the farthest advance of a glacier.
Terrace—Relatively flat or inclined surface bounded by an ascending slope and a steep descending slope on opposite sides. Along the Atlantic Coast there are the Okefenokee, Pamlico, Silver Bluff, and Wicomico Terraces, all ancient.
Terraced—Step-like terraces in series.
Tidal Flat—Marshy or muddy area which is covered and uncovered by the rise and fall of the tide.
Tidal Marsh or Swamp—Similar to a tidal flat but generally filled with reeds, etc.
Tidal Meanders—Meanders of channels in tidal flats and affected by the rise and fall of tides.
Till Plain—Plain on which sediment carried by a glacier has been deposited.
Thermal Spring—Hot spring.
Tombolo—Bar connecting an island or a large rock with the mainland or with another island.
Topographic Grain—Uniform patterns of drainage and ridges.
Topographic Map—Map showing the physical features of a land surface, generally by means of contours and drainage lines. (*See* the discussion in the Introduction.)
Topographic Map Symbols—See chart on page 135.
Topography—Shape, relief, elevation, and drainage of the land. (*See* the discussion in the Introduction.)
Trellis Drainage—A stream pattern in which master and tributary streams are at right angles or nearly so.
Truncated—Terminated abruptly as a spur ridge in a valley cut off by glacial action. *See* U-shaped Valley.

Undercut Slope—Very steep bluff at the meander of a river or stream caused by the downward erosion force of the stream; occurs opposite the slip-off slope.
Uplift—Elevation of any extensive part of the Earth's surface relative to adjacent parts.
Urban Area—Densely built area of more than three-fourths of a square mile which is generally shown on a topographic map by a tinted overprint. Only the landmark buildings are shown. Elsewhere on a quadrangle, buildings are shown by rectangular symbols, to scale when possible.
U-Shaped Valley—Valley so completely eroded by glacial action that it is U-shaped and even the original truncated spurs are eliminated.
USMM—United States Mineral Monument to which mining claims are referenced.

Valley—Low-lying land, generally bounded by hills or mountain ranges and usually traversed by a river or watercourse which receives the drainage of the surrounding heights.
Valley Sink—*See* Blind Valley.
Vly—Small swamp or marsh; valley where water collects.
Volcanic Cone—Cone-shaped hill or mountain formed by volcanic discharges.
Volcanic Crater—*See* Crater.
Volcanic Crater or Cone, Breached—*See* Breached Crater.
Volcanic Neck or Plug—Solidified material filling a vent of a dead volcano; stands as a crag or tower of igneous rock when high and alone.
Volcano—Vent in the Earth's crust from which molten lava issues, frequently building up a mountain.
V-Shaped Valley—Gorge or valley with evenly sloping sides.
Vulcanism—Result of volcanic action.

Warm Spring—Thermal spring.
Wash—Intermittent stream filled with coarse sediment and confined to semi-arid regions.
Waterfall—Fall or falls.
Water Gap—Gap or low depression in a mountain ridge through which a stream flows.
Wave-Cut Cliff—Cliff formed by wave action.
Wind Gap—Low notch or gap in a ridge where a stream formerly flowed.

TOPOGRAPHIC MAP SYMBOLS

VARIATIONS WILL BE FOUND ON OLDER MAPS

Hard surface, heavy duty road, four or more lanes	
Hard surface, heavy duty road, two or three lanes	
Hard surface, medium duty road, four or more lanes	
Hard surface, medium duty road, two or three lanes	
Improved light duty road	
Unimproved dirt road and trail	
Dual highway, dividing strip 25 feet or less	
Dual highway, dividing strip exceeding 25 feet	
Road under construction	

Railroad, single track and multiple track	
Railroads in juxtaposition	
Narrow gage, single track and multiple track	
Railroad in street and carline	
Bridge, road and railroad	
Drawbridge, road and railroad	
Footbridge	
Tunnel, road and railroad	
Overpass and underpass	
Important small masonry or earth dam	
Dam with lock	
Dam with road	
Canal with lock	

Buildings (dwelling, place of employment, etc.)	
School, church, and cemetery	⌐⌐⌐ † Cem
Buildings (barn, warehouse, etc.)	
Power transmission line	
Telephone line, pipeline, etc. (labeled as to type)	
Wells other than water (labeled as to type)	o Oil ... o Gas
Tanks; oil, water, etc. (labeled as to type)	• • ● ⊘ Water
Located or landmark object; windmill	o ... ⚒
Open pit, mine, or quarry; prospect	✕ ... x
Shaft and tunnel entrance	■ ... Y

Horizontal and vertical control station:
    Tablet, spirit level elevation ..... BM △5653

Other recoverable mark, spirit level elevation	△5455
Horizontal control station: tablet, vertical angle elevation	VABM △9519
Any recoverable mark, vertical angle or checked elevation	△3775
Vertical control station: tablet, spirit level elevation	BM ✕957
Other recoverable mark, spirit level elevation	✕954
Checked spot elevation	✕4675
Unchecked spot elevation and water elevation	✕5657 870

Boundary, national	
State	
County, parish, municipio	
Civil township, precinct, town, barrio	
Incorporated city, village, town, hamlet	
Reservation, national or state	
Small park, cemetery, airport, etc.	

Index contour		Intermediate contour	
Supplementary contour		Depression contours	
Fill		Cut	
Levee		Levee with road	
Mine dump		Wash	
Tailings		Tailings pond	
Strip mine		Distorted surface	
Sand area		Gravel beach	

Perennial streams		Intermittent streams	
Elevated aqueduct		Aqueduct tunnel	
Water well and spring	o ... o~	Disappearing stream	
Small rapids		Small falls	
Large rapids		Large falls	
Intermittent lake		Dry lake	
Foreshore flat		Rock or coral reef	
Sounding, depth curve	10	Piling or dolphin	o
Exposed wreck		Sunken wreck	
Rock, bare or awash; dangerous to navigation			

Marsh (swamp) .....   Submerged marsh .....